Computational Modeling in Bioinformatics

Computational Modeling in Bioinformatics

Editor: Alexis White

www.statesacademicpress.com

States Academic Press,
109 South 5th Street,
Brooklyn, NY 11249, USA

Visit us on the World Wide Web at:
www.statesacademicpress.com

© States Academic Press, 2022

This book contains information obtained from authentic and highly regarded sources. Copyright for all individual chapters remain with the respective authors as indicated. All chapters are published with permission under the Creative Commons Attribution License or equivalent. A wide variety of references are listed. Permission and sources are indicated; for detailed attributions, please refer to the permissions page and list of contributors. Reasonable efforts have been made to publish reliable data and information, but the authors, editors and publisher cannot assume any responsibility for the validity of all materials or the consequences of their use.

ISBN: 978-1-63989-123-8 (Hardback)

Trademark Notice: Registered trademark of products or corporate names are used only for explanation and identification without intent to infringe.

Cataloging-in-Publication Data

Computational modeling in bioinformatics / edited by Alexis White.
　p. cm.
Includes bibliographical references and index.
ISBN 978-1-63989-123-8
1. Bioinformatics. 2. Computer simulation. 3. Computational biology.
4. Systems biology. I. White, Alexis.

QH324.2 .B56 2022
570.285--dc23

Table of Contents

	Preface VII
Chapter 1	**Bioinformatics Applied to Proteomics**..1 Simone Cristoni and Silvia Mazzuca
Chapter 2	**Protein Progressive MSA Using 2-Opt Method**............................... 27 Gamil Abdel-Azim, Aboubekeur Hamdi-Cherif, Mohamed Ben Othman and Z. A. Aboeleneen
Chapter 3	**Evolutionary Bioinformatics with a Scientific Computing Environment**.. 43 James J. Cai
Chapter 4	**Parallel Processing of Complex Biomolecular Information: Combining Experimental and Computational Approaches** .... 67 Jestin Jean-Luc and Lafaye Pierre
Chapter 5	**Strengths and Weaknesses of Selected Modeling Methods Used in Systems Biology**... 89 Pascal Kahlem et al.
Chapter 6	**Signal Processing Methods for Capillary Electrophoresis** ....111 Robert Stewart, Iftah Gideoni and Yonggang Zhu
Chapter 7	**Understanding Tools and Techniques in Protein Structure Prediction**... 135 Geraldine Sandana Mala John, Chellan Rose and Satoru Takeuchi
Chapter 8	**Systematic and Phylogenetic Analysis of the Ole e 1 Pollen Protein Family Members in Plants** .... 163 José Carlos Jiménez-López, María Isabel Rodríguez-García and Juan de Dios Alché

Chapter 9 **The Information Systems for DNA Barcode Data**..........................179
        Di Liu and Juncai Ma

Chapter 10 **Biological Data Modelling and Scripting in R**................................201
        Srinivasan Ramachandran et al.

Permissions

List of Contributors

Index

Preface

This book has been an outcome of determined endeavour from a group of educationists in the field. The primary objective was to involve a broad spectrum of professionals from diverse cultural background involved in the field for developing new researches. The book not only targets students but also scholars pursuing higher research for further enhancement of the theoretical and practical applications of the subject.

Bioinformatics is an inter-disciplinary field which is concerned with the development of new methods for understanding large and complex biological data structures. It combines the principles of biology, mathematics, computer science, statistics and information engineering for the analysis of biological molecules such as DNA and RNA. A computational model is a complex non-linear model which requires extensive resources to study a system using computer simulation. Bioinformatics makes use of computational modeling systems to study DNA sequencing, genome annotation, genomics and mutation. It also finds applications in the field of gene expression, protein expression and analysis of cellular organization. The book studies, analyses and upholds the pillars of bioinformatics and computational modeling, and its utmost significance in modern times. The various studies that are constantly contributing towards advancing technologies and evolution of this field are examined in detail. This book attempts to assist those with a goal of delving into this field.

It was an honour to edit such a profound book and also a challenging task to compile and examine all the relevant data for accuracy and originality. I wish to acknowledge the efforts of the contributors for submitting such brilliant and diverse chapters in the field and for endlessly working for the completion of the book. Last, but not the least; I thank my family for being a constant source of support in all my research endeavours.

Editor

Bioinformatics Applied to Proteomics

Simone Cristoni[1] and Silvia Mazzuca[2]
[1]Ion Source Biotechnologies srl, Milano,
[2]Plant Cell Physiology laboratory, Università della Calabria, Rende,
Italy

1. Introduction

Proteomics is a fundamental science in which many sciences in the world are directing their efforts. The proteins play a key role in the biological function and their studies make possible to understand the mechanisms that occur in many biological events (human or animal diseases, factor that influence plant and bacterial grown). Due to the complexity of the investigation approach that involve various technologies, a high amount of data are produced. In fact, proteomics has known a strong evolution and now we are in a phase of unparalleled growth that is reflected by the amount of data generated from each experiment. That approach has provided, for the first time, unprecedented opportunities to address biology of humans, animals, plants as well as micro-organisms at system level. Bioinformatics applied to proteomics offered the management, data elaboration and integration of these huge amount of data. It is with this philosophy that this chapter was born.
Thus, the role of bioinformatics is fundamental in order to reduce the analysis time and to provide statistically significant results. To process data efficiently, new software packages and algorithms are continuously being developed to improve protein identification, characterization and quantification in terms of high-throughput and statistical accuracy. However, many limitations exist concerning bioinformatic spectral data elaboration. In particular, for the analysis of plant proteins extensive data elaboration is necessary due to the lack of structural information in the proteomic and genomic public databases. The main focus of this chapter is to describe in detail the status of bioinformatics applied to proteomic studies. Moreover, the elaboration strategies and algorithms that have been adopted to overcome the well known limitations of the protein analysis without database structural information are described and disclosed.
This chapter will get rid of light on recent developments in bioinformatic and data-mining approaches, and their limitations when applied to proteomic data sets, in order to reinforce the interdependence between proteomic technologies and bioinformatics tools. Proteomic studies involve the identification as well as qualitative and quantitative comparison of proteins expressed under different conditions, together with description of their properties and functions, usually in a large-scale, high-throughput format. The high dimensionality of data generated from these studies will require the development of improved bioinformatics tools and data-mining approaches for efficient and accurate data analysis of various

biological systems (for reviews see, Li et al, 2009; Matthiesen & Jensen, 2008; Wright et al, 2009). After a rapid moving on the wide theme of the genomic and proteomic sciences, in which bioinformatics find their wider applications for the studies of biological systems, the chapter will focus on mass spectrometry that has become the prominent analytical method for the study of proteins and proteomes in post-genome era. The high volumes of complex spectra and data generated from such experiments represent new challenges for the field of bioinformatics. The past decade has seen an explosion of informatics tools targeted towards the processing, analysis, storage, and integration of mass spectrometry based proteomic data. In this chapter, some of the more recent developments in proteome informatics will be discussed. This includes new tools for predicting the properties of proteins and peptides which can be exploited in experimental proteomic design, and tools for the identification of peptides and proteins from their mass spectra. Similarly, informatics approaches are required for the move towards quantitative proteomics which are also briefly discussed. Finally, the growing number of proteomic data repositories and emerging data standards developed for the field are highlighted. These tools and technologies point the way towards the next phase of experimental proteomic and informatics challenges that the proteomics community will face.

The majority of the chapter is devoted to the description of bioinformatics technologies (hardware and data management and applications) with particular emphasis on the bioinformatics improvements that have made possible to obtain significant results in the study of proteomics. Particular attention is focused on the emerging statistic semantic, network learning technologies and data sharing that is the essential core of system biology data elaboration.

Finally, many examples of bioinformatics applied to biological systems are distributed along the different section of the chapter so to lead the reader to completely fill and understand the benefits of bioinformatics applied to system biology.

2. Genomics versus proteomics

There have been two major diversification paths appeared in the development of bioinformatics in terms of project concepts and organization, the -omics and the bio-. These two historically reflect the general trend of modern biology. One is to go into molecular level resolution. As one of the -omics and bio- proponents, the -omics trend is one of the most important conceptual revolutions in science. Genetic, microbiology, mycology and agriculture became effectively molecular biology since 1970s. At the same time, these fields are now absorbing omics approach to understand their problems more as complex systems. Omics is a general term for a broad discipline of science and engineering for analyzing the interactions of biological information objects in various omes. These include genome, proteome, metabolome, expressome, and interactome. The main focus is on mapping information objects such as genes, proteins, and ligands finding interaction relationships among the objects, engineering the networks and objects to understand and manipulate the regulatory mechanisms and integrating various omes and omics subfields.

This was often done by researchers who have taken up the large scale data analysis and holistic way of solving bio-problems. However, the flood of such -omics trends did not occur until late 1990s. Until that time, it was by a relatively small number of informatics advanced people in Europe and the USA. They included Medical Research Council [MRC] Cambridge, Sanger centre, European Bioinformatics Institute [EBI], European Molecular

Biology Laboratory [EMBL], Harvard, Stanford and others. We could clearly see some people took up the underlying idea of -ome(s) and -omics quickly, as biology was heading for a more holistic approach in understanding the mechanism of life. Whether the suffix is linguistically correct or not, the -omics suffix changed in the way many biologists view their research activity. The most profound one is that biologists became freshly aware of the fact that biology is an information science more than they have thought before.

In general terms, genomics is the -omics science that deals with the discovery and noting of all the sequences in the entire genome of a particular organism. The genome can be defined as the complete set of genes inside a cell. Genomics, is, therefore, the study of the genetic make-up of organisms. Determining the genomic sequence, however, is only the beginning of genomics. Once this is done, the genomic sequence is used to study the function of the numerous genes (functional genomics), to compare the genes in one organism with those of another (comparative genomics), or to generate the 3-D structure of one or more proteins from each protein family, thus offering clues to their function (structural genomics). At today a list of sequenced eukaryotic genomes contains all the eukaryotes known to have publicly available complete nuclear and organelle genome sequences that have been assembled, annotated and published. Starting from the first eukaryote organism *Saccharomyces cerevisiae* to have its genome completely sequenced at 1998, further genomes from 131 eukaryotic organisms were released at today. Among them 33 are Protists, 16 are Higher plants, 26 are Fungi, 17 are Mammals Humans included, 9 are non-mammal animals ,10 are Insects, 4 Nematodes, remaining 11 genomes are from other animals and as we write this chapter, others are still to be sequenced and will be published during the editing of this book. A special note should be paid to the efforts of several research teams around the world for the sequencing of more than 284 different Eubacteria, whose numbers increased by 2-3% if we consider the sequencing of different strains for a single species; also a list of sequenced archaeal genomes contains 28 Archeobacteria known to have available complete genome sequences that have been assembled, annotated and deposited in public databases.

A striking example of the power of this kind of -omics and knowledge that it reveals is that the full sequencing of the human genome has dramatically accelerated biomedical research and diagnosis forecast; very recently Eric S. Lander (2011) explored its impact, in the decade since its publication, on our understanding of the biological functions encoded in the human genome, on the biological basis of inherited diseases and cancer, and on the evolution and history of the human species; also he foresaw the road ahead in fulfilling the promise of genomics for medicine.

In the other side of living kingdoms, genomics and biotechnology are also the modern tools for understanding plant behavior at the various biological and environmental levels. In The Arabidopsis Information Resource [TAIR] a continuously updated database of genetic and molecular biology data for the model higher plant *Arabidopsis thaliana* is maintained (TAIR Database, 2009)

This data available from TAIR include the complete genome sequence along with gene structure, gene product information, metabolism, gene expression, DNA and seed stocks, genome maps, genetic and physical markers, publications, and information about the *Arabidopsis* research community. Gene product function data is updated every two weeks from the latest published research literature and community data submissions. Gene structures are updated 1-2 times per year using computational and manual methods as well as community submissions of new and updated genes.

Genomics provides also boosting to classical plant breeding techniques, well summarized in the Plants for the Future technology platform (http://www.epsoweb.eu/catalog/tp/tpcom_home.htm). A selection of novel technologies come out that are now permitting researchers to identify the genetic background of crop improvement, explicitly the genes that contribute to the improved productivity and quality of modern crop varieties. The genetic modification (GM) of plants is not the only technology in the toolbox of modern plant biotechnologies. Application of these technologies will substantially improve plant breeding, farming and food processing. In particular, the new technologies will enhance the ability to improve crops further and, not only will make them more traceable, but also will enable different varieties to exist side by side, enhancing the consumer's freedom to choose between conventional, organic and GM food. In these contexts agronomical important genes may be identified and targeted to produce more nourishing and safe food; proteomics can provide information on the expression of transgenic proteins and their interactions within the cellular metabolism that affects the quality, healthy and safety of food. Taking advantage of the genetic diversity of plants will not only give consumers a wider choice of food, but it will also expand the range of plant derived products, including novel forms of pharmaceuticals, biodegradable plastics, bio-energy, paper, and more. In this view, plant genomics and biotechnology could potentially transform agriculture into a more knowledge-based business to address a number of socio-economic challenges.

In systems biology (evolutionary and/or functionally) a central challenge of genomics is to identify genes underlying important traits and describe the fitness consequences of variation at these loci (Stinchcombe et al., 2008). We do not intend to give a comprehensive overview of all available methods and technical advances potentially useful for identifying functional DNA polymorphisms, but rather we explore briefly some of promising recent developments of genomic tools from which proteomics taken its rise during the last twenty years, applicable also to non model organisms.

The genome scan, became one of the most promising molecular genetics (Oetjen et al., 2010). Genome scans use a large number of molecular markers coupled with statistical tests in order to identify genetic loci influenced by selection (Stinchombe & Hoekstra, 2008). This approach is based on the concept of 'genetic hitch-hiking' (Maynard Smith & Haigh, 1974) that predicts that when neutral molecular markers are physically linked to functionally important and polymorphic genes, divergent selection acting on such genes also affects the flanking neutral variation. By genotyping large numbers of markers in sets of individuals taken from one or more populations or species, it is possible to identify genomic regions or 'outlier loci' that exhibit patterns of variation that deviate from the rest of the genome due to the effects of selection or treats (Vasemägi & Primmer 2005). An efficient way of increasing the reliability of genome scans, which does not depend on the information of the genomic location of the markers, is to exploit polymorphisms tightly linked to the coding sequences, such as expressed sequence tag (EST) linked microsatellites (Vigouroux et al., 2002; Vasemägi et al., 2005). Because simple repeat sequences can serve as promoter binding sites, some microsatellite polymorphisms directly upstream of genes may have a direct functional significance (Li et al., 2004).

EST libraries represent sequence collections of all mRNA (converted into complementary or cDNA) that is transcribed at a given point in time in a specific tissue (Bouck & Vision, 2007). EST libraries have been constructed and are currently being analyzed for many species whose genomes are not completed. EST library also provide the sequence data for expression analysis using *Quantitative real-time PCR (QPCR)*, as well as for transcription profiling using *microarrays* and, finally, the EST database can be a valuable tool for identifying new candidate polymorphism in proteins of specific interest. QPCR is a method that can measure the abundance of mRNA (converted in cDNA) of specific genes (Heid et al., 1996). The expression of a target gene can be related to the total RNA input, or it can be quantified in relation to the expression of a reference gene, the housekeeping gene (HKG, i.e. gene always expressed at the same level). Unfortunately, a universal reference gene that is expressed uniformly, in all biological conditions in all tissues, does not exist. For each experimental setup using QPCR, the choice of HKG must reflect the tissue used and the experimental treatment.

While QPCR can only handle a few candidate genes, *microrrays technology* quantifies the expression level of hundreds to thousands of genes simultaneously, providing a powerful approach for the analysis of global transcriptional response (Yauk & Berndt, 2007). For example, the analysis of mRNA via genomic arrays is one approach to finding the genes differentially expressed across two kind of tissue or sample obtained under two experimental conditions or to finding the genes that matter to organisms undergoing environmental stress. Additionally, microarray data can be used to distinguish between neutral and adaptive evolutionary processes affecting gene expression (e.g. Gibson, 2002; Feder & Walser, 2005; Whitehead & Crawford, 2006). Nevertheless, a sequencing revolution is currently driven by new technologies, collectively referred to as either 'next-generation' sequencing, 'highthroughput' sequencing, 'ultra-deep' sequencing or 'massively parallel' sequencing. These technologies allow us the large scale generation of ESTs efficiently and cost-effectively available at the National Centre Biotechnology Information database [NCBI-dbEST] (http://www.ncbi.nlm.nih.gov/dbEST); Shendure et al., 2005). There are increasing studies in which 454 technologies, combined or not with Solexa/Illumina, are used to characterize transcriptomes in several plant and animal species (Emrich et al., 2007; Metzker, 2010; Eveland et al., 2008; Bellin et al., 2009). To give an idea of the potential implications of these sequencing technologies it is enough to know that the pyrosequencing delivers the microbial genome sequence in 1 hour, thus upsetting perspectives in basic research, phylogenetic analysis, diagnostics as in industrial applications (Clarke, 2005; Hamady et al., 2010; Yang et al., 2010; Claesson et al., 2009). Even in full sequenced genomes, such as in *Arabidopsis* or humans, this deep sequencing is allowing to identify new transcripts not present in previous ESTs collections (Weber et al., 2007; Sultan et al., 2010). Also specific transcriptomes are being generated in species for which previous genomic resources lacked because of the large size of their genomes (Alagna et al., 2009; Wang et al., 2009; Craft et al., 2010) The new transcripts are also being used for microarrays design (Bellin et al.,2009), and also for high throughput SSRs or SNPs identification. SNP detection is performed by aligning raw reads from different genotypes to a reference genome or transcriptome previously available in plants (Barbazuk et al., 2006), as in plants, (Trick et al., 2009; Guo et al., 2010), animals (Satkoski et al., 2008) and humans (Nilsson et al., 2004).

De novo assembly of raw sequences coming from a set of genotypes, followed by pairwise comparison of the overlapping assembled reads has also successfully used in species lacking any significant genomic or transcriptomic resources (Novaes et al., 2008). The principle behind these applications (as termed sequence census methods) is simple: complex DNA or RNA samples are directly sequenced to determine their content, without the requirement for DNA cloning. Thus, these novel technologies allow the direct and cost-effective sequencing of complex samples at unprecedented scale and speed, making feasible to sequence not only genomes, but also entire transcriptomes expressed under different conditions. Moreover, a unique feature of sequence census technologies is their ability to identify, without prior knowledge, spliced transcripts by detecting the presence of sequence reads spanning exon-exon junctions. Hence, next-generation sequencing delivers much more information at affordable costs, which will increasingly supersede microarray based approaches (Marguerat et al., 2008).

It is noteworthy, however, that transcription profiling has been questioned as an effective tool for the discovery of genes that are functionally important and display variable expression (e.g. Feder & Walser, 2005). In fact, the vast majority of genes implicated by transcriptomics can be expected to have no phenotype. Furthermore, even if the synthesis of mature protein is closely linked to the abundance of its corresponding mRNA, the concentration of mature protein is the net of its synthesis and degradation. Degradation mechanisms and rates can vary substantially and lead to corresponding variation in protein abundance (Feder & Walser, 2005). The physiological measurements of protein abundance for selected gene candidate could be a valuable addition to pure transcriptomic studies (Jovanovic et al., 2010).

It is reasonable that a method should measure the most relevant output of gene expression, namely dependent changes in protein amounts from potential target genes. Moreover, to be worthwhile, the method should be easy to use, fast, sensitive, reproducible, quantitative and scalable, as several hundred proteins have to be tested. A technique that promises to fulfill most of those criteria is proteomics which is experiencing considerable progress after the massive sequencing of many genomes from yeast to humans for both basic biology and clinical research (Tyers & Mann, 2003). For identifying and understanding the proteins and their functions from a cell to a whole organism, proteomics is a necessity in the assortment of –omics technologies.

Historically, the term *proteome* was coined by Mark Wilkins first in 1994 as a blend of proteins and genome and Wilkins used it to describe the entire complement of proteins expressed by a genome, cell, tissue or organism. Subsequently this term has been specified to contain all the expressed proteins at a given time point under defined conditions and it has been applied to several different types of biological systems (Doyle, 2011; Ioannidis, 2010; Heazlewood, 2011; Prokopi & Mayr, 2011; Wienkoop et al, 2010).

In a basic view, a cellular proteome is the collection of proteins found in a particular cell type under a particular set of conditions such as differentiation stage, exposure to hormone stimulation inside tissues or changing of physical parameters in an environment. It can also be useful to consider an organism's complete proteome, which can be conceptualized as the complete set of proteins from all of the various cellular proteomes. This is very roughly the protein equivalent of the genome. The term "proteome" has also been used to refer to the collection of proteins in certain sub-cellular biological systems. For example, all of the proteins in a virus can be called a viral proteome. The proteome is larger than the genome, especially in eukaryotes, in the sense that there are more proteins than genes. This is due to alternative splicing of genes and post-translational modifications like glycosylation or phosphorylation. Moreover the proteome has at least two levels of complexity lacking in the genome. When the genome is defined by the sequence of nucleotides, the proteome cannot be limited to the sum of the sequences of the proteins present. Knowledge of the proteome requires knowledge of the structure of the proteins in the proteome and the functional interaction between the proteins.

The escalating sequencing of genomes and the development of large EST databases have provided genomic bases to explore the diversity, cellular evolution and adaption ability of organisms. However, by themselves, these data are of limited use when they try to fully understand processes such as development, physiology and environmental adaptation. Taking advances from genomic information the proteomics can assign function to proteins and elucidate the related metabolism in which the proteins act (Costenoble et al., 2011; Chen et al., 2010; Joyard et al., 2010, Tweedie-Cullen & Mansuy, 2010).

In a wide-ranging functional view, proteomics is matching to genomics: through the use of pure genome sequences, open reading frames (ORFs) can be predicted, but they cannot be used to determine if or when transcription takes place. Proteomics, indicating at what level a protein is expressed, can also provide information about the conditions under which a protein might be expressed, its cellular location (Agrawal et al., 2010; Jamet et al., 2006; Rossignol et al., 2006; Tyers & Mann, 2003), the relative quantities (Yao et al., 2001; Molloy et al., 2005), and what protein–protein interactions take place (Giot et al., 2003; Schweitzer et al., 2003). Genomics, in essence, demonstrates which genes are involved, whereas proteomics can show clearer relationships by illustrating functional similarities and phenotypic variances.

Because the environments in which organisms live is dynamic, the success of a species depends on its ability to rapidly adapt to varying limiting factors such as light (for plants above all), temperature, diet or nutrient sources. Since the proteome of each living cell is dynamic, proteomics allows investigators to clarify if and to what extent various pathways are utilized under varying conditions, triggered by the action of the environment on the system, and relative protein-level response times. In other words how organisms are able to biochemically survive to conditions imposed by environment.

Huge amount of data have been accumulated and organized in world-wide web sites served for proteomics as main proteomics-related web sites have been lunched (Tab 1).

For example The Human Protein Reference Database represents a centralized platform to visually illustrate and integrate information pertaining to domain architecture, post-translational modifications, interaction networks and disease association for each protein in the human proteome; on the ExPASy Proteomics site, tools are available locally to the server or are developed and hosted on other servers.

As concerning plant proteomics, the research community is well served by a number of online proteomics resources that hold an abundance of functional information. Recently, members of the *Arabidopsis* proteomics community involved in developing many of these resources decided to develop a summary aggregation portal that is capable of retrieving proteomics data from a series of online resources (Joshi et al., 2010, http://gator.masc-proteomics.org/). This means that information is always up to date and displays the latest datasets. The site also provides hyperlinks back to the source information hosted at each of the curated databases to facilitate analysis of the primary data. Deep analyses have also performed on organelle proteomics as in protists, animals and plants. A well-known database, launched in 2004, is devoted to proteomics of mitochondria in yeast (Ohlmeier et al., 2004; http://www.biochem.oulu.fi/proteomics/ymp.html), while the Nuclear Protein Database [NPD] is a curated database that contains information on more than 1300

\multicolumn{3}{c}{**World-wide Web Sites served for Proteomics**}		
Name	Web site	Characteristics
WORLD-2DPAGE Index to federated 2-D PAGE database	http://www.expasy.ch/ch2d/2d-index.htm	integrated proteome database for use in cancer research by two-dimensional difference gel electrophoresis (2D-DIGE)
2D GEL DATABASES WORLD-WIDE (GeMDBJ Database Link Station)	https://gemdbj.nibio.go.jp/dgdb/dige/servlet/DigeLinkTo2dDatabaseCountryServlet	
ExPASy Proteomics tools	http://www.expasy.ch/tools/	Protein identification and characterization with peptide mass fingerprinting data
Mascot Search	http://www.matrixscience.com/search_form_select.html	search engine which uses mass spectrometry data to identify proteins from primary sequence databases
ProteinProspector	http://prospector.ucsf.edu/prospector/mshome.htm	Proteomics tools for mining sequence databases in conjunction with Mass Spectrometry experiments.
MASCP Gator	http://gator.masc-proteomics.org/ MASCP Gator	the portal provides hyperlinks back to the source information hosted at each of the curated databases
PPDB	http://ppdb.tc.cornell.edu/ The Plant Proteome Database	database dedicated to the whole plant proteome
AT_Chloro	http://www.grenoble.prabi.fr/at_chloro/ AT_CHLORO Database	database dedicated to the chloroplast proteome from Arabidopsis thaliana
\multicolumn{3}{c}{**Main Proteomics-related Web Sites**}		
The Japanese Electrophoresis Society Home Page	http://www.jes1950.jp/english/	it promotes the development of electrophoretic technologies and their applications
DNA Data Bank of Japan (DDBJ)	http://www.ddbj.nig.ac.jp/	is the sole nucleotide sequence data bank in Asia, which is officially certified to collect nucleotide sequences from researchers
HUPO (Human Proteome Organisation)	http://www.hupo.org/	international scientific organization representing and promoting proteomics through international cooperation and collaborations
Electrophoresis (Wiley-VCH)	http://www.wiley-vch.de/publish/en/journals/alphabeticIndex/2027/	is one of the world's leading journals for new analytical and preparative methods and for innovative applications on all aspects of electrophoresis
UniProtKB/Swiss-Prot Protein Knowledgebase	http://us.expasy.org/sprot/relnotes/spwrnew.html	provide comprehensive and non-redundant complete proteome sets for all species that are currently covered

Table 1. List of the main proteome databases.

vertebrate proteins that are thought, or are known, to localize to the cell nucleus. The database can be accessed at http://npd.hgu.mrc.ac.uk and is updated monthly. Very recently, plant organelle proteomics has experienced a rapid growth in the field of functional proteomics (see the review, Agrawal et al., 2010); from this efforts gave rise seven main websites of which two are devoted to the plastid (Plant Proteomic DataBase [PPDB] http://ppdb.tc.cornell.edu/), two are specific for mitochondria (Arabidopsis Mitochondrial Protein DataBase [AMPDB], http://plantenergy.uwa.edu.au/application/ampdb/; Arabidopsis Mitochondrial Protein Project [AMPP], http://gartenbau.unihannover.de/genetic/AMPP), one is an accurate database of comprehensive chloroplast proteome (AT_Chloro, http://www.grenoble.prabi.fr/protehome/grenoble-plant-proteomics/).

An area of study within proteomics is 'expression proteomics', which is defined as the use of quantitative protein-level measurements of gene expression to characterize biological processes and deduce the mechanisms of gene expression control. Expression proteomics allows researchers to obtain a quantitative description of protein expression and its changes under the influence of biological perturbations, the occurrence of post-translational modifications and the distribution of specific proteins within cells (Baginsky et al., 2010; Roth et al., 2010).

As an example of high technological potential of expression proteomics, in the last ten years plant proteomics research has been conducted in several land species achieving a high degree of knowledge of the dynamics of the proteome in many model plants (Agrawal & Rakwal, 2005; Baerenfaller et al., 2008; Grimplet et al., 2009; Komatsu, 2008; Plomion et al., 2006) and thereafter translating this knowledge in other species whose genome sequence is still under construction. The most successful studies are those which use separation of subcellular compartments (Haynes & Roberts, 2007; Dunkley et al., 2006; Agrawal et al., 2010) such as mitochondria (Heazlewood et al., 2005), chloroplast (Ferro et al., 2010), endoplasmic reticulum (Maltman et al., 2007), peroxisomes (Fukao et al., 2002), plastoglobules (Grennan, 2008), vacuoles (Jaquinod et al., 2007), nucleus (Repetto et al., 2008) since they contain a limited number of proteins thus helping the protein identification. Since 30 years, the greater part of research into the plant proteome has utilized two-dimensional sodium dodecyl sulphate–polyacrylamide gel electrophoresis (2D SDS–PAGE) for the protein separation step, which is usually followed by protein identification by mass spectrometry (MS). Proteomics, the study of the proteome, has largely been practiced through the separation of proteins by two dimensional gel electrophoresis. In the first dimension, the proteins are separated by isoelectric focusing, which resolves proteins on the basis of charge. In the second dimension, proteins are separated by molecular weight using SDS-PAGE. The gel is dyed to visualize the proteins and the spots on the gel are proteins that have migrated to specific locations.

The number of spots resolved in plant proteomics 2D projects depends on the chosen tissue and plant species as well as the protein nature (i.e. basic or acid, soluble or membrane-associated; Tsugita & Kamo, 1994; Porubleva et al., 2001). The gel plugs, containing the proteins of interest are collected to further analyses by mass MS approaches and database searches (Chevalier, 2010; Yates *et al.*, 2009; Zhao &d Lin, 2010). This method where proteins are analyzed after enzymatic digestion is widely used for high complexity samples in large scale analyses and it is known as "bottom up approach" that was discussed in detail in the next paragraph. Attention must given to the importance of sound statistical treatment of the resultant quantifications in the search for differential expression. Despite wide availability of proteomics software, a number of challenges have yet to be overcome regarding algorithm accuracy, objectivity and automation, generally due to deterministic spot-centric approaches that discard information early in the pipeline, propagating errors. We review recent advances in signal and image analysis algorithms in 2-DE, MS, LC/MS and Imaging MS.

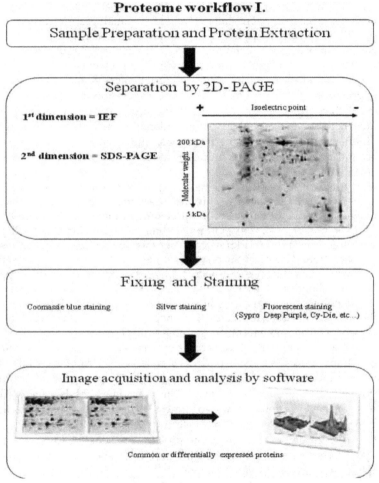

Fig. 1. Proteome workflow I: after sample preparation and protein extraction, proteins are initially separated by isoelectric focusing (IEF) in which they migrate along an IEF strip which has a pH gradient between a cathode and an anode; the migration of each protein ends when it reaches its isoelectric point in the gradient. This strip is then applied to a SDS polyacrylamide gel in which the second dimension of the separation occurs according to molecular weights. After fixation, the gel is stained by different techniques and its digital image is acquired to be further analyzed by specific softwares, in order to found the significant differentially expressed proteins.
With permission of Nova Science Publishers, Inc.

3. Bioinformatics in proteomics

Mass spectrometry became a very important tool in proteomics: it has made rapid progresses as an analytical technique, particularly over the last decade, with many new types of hardware being introduced (Molloy et al, 2005; Matthiesen and Jensen, 2008, Yates et al, 2009). Moreover, constant improvements have increased the levels of MS sensitivity, selectivity as well as mass measurement accuracy. The principles of mass spectrometry can be envisaged by the following four functions of the mass spectrometer: i) peptide ionization; ii) peptide ions analyses according to their mass/charge ratio (m/z) values ; iii) acquisition of ion mass data ; iv) measurement of relative ion abundance. Ionization is fundamental as the physics of MS relies upon the molecule of interest being charged, resulting in the formation of positive ions, and, depending on the ionization method, fragment ions. These ion species are visualized according to their corresponding m/z ratio(s), and their masses assigned. Finally, the measurement of relative ion abundance, based on either peak height or peak area of sample(s) and internal standard(s), leads to a semi-quantitative request.

3.1 Typical procedure for proteome analysis

Proteome data elaboration procedure is different depending of the study target. In general the studies can be qualitative in order to characterize the organisms expressed proteome and quantitative to detect potential biomarker related to disease or other organism proprieties. The principal proteomics studies are:
i. Full proteomics (qualitative);
ii. Functional proteomics (relative quantitation studies);
iii. Post translational modification functional proteomics (qualitative and relative quantitation studies)

3.2 Data elaboration for full proteome analysis

In full proteomics analysis (Armengaud et al. 2010) the proteins are usually extracted and qualitatively identified. These studies are usually performed in order to understand what proteins are expressed by the genome of the organism of interest. The general analytical scheme is reported in Figure 2.

Basically, after protein separation, mainly through gel electrophoresis or other separation approaches (liquid chromatography etc.), proteins are identified by means of mass spectrometric technique. Two kind of data processing algorithms can be employed depending by the analytical technology used to analyze the proteins. The two approaches are:
i. Bottom up approach. It is used to identify the protein of interest after enzymatic or chemical digestion;
ii. Top down approach. In this case proteins are not digested but directly analyzed by mass spectrometric approaches;

In the former case (bottom up) the protein are digested by means of enzymatic or chemical reaction and the specific peptides produced are then analyzed to identify the protein of interest. This results can be obtained using mass spectrometric mass analyzer that can operate in two conditions: a) full scan peptide mass fingerprint (MS) and b) tandem mass spectrometry (MS/MS). In the case a) the mass/charge (m/z) ratio of the peptide is obtained using high resolution and mass accurate analyzer (time of flight, FTICR; see

Cristoni et al., 2004, 2003). The combination of the high accurate m/z ratio of the detected peptides is checked against the theoretical one generated by virtual digestion of the proteins present in the known database. A list of protein candidates is so obtained with relative statistical identification score, correlated to the number of peptides detected, per proteins and peptide mass accuracy. The principal software package used for this kind of data elaboration are reported in table 2.

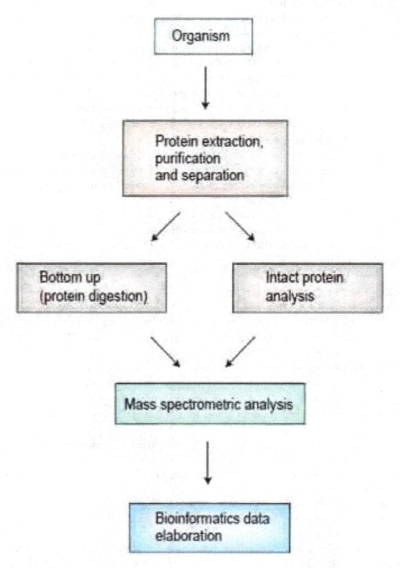

Fig. 2. General analytical scheme of Full proteomic analysis.

Software name	Availability	Web	Categor
Aldent	fre	http://www.expasy.org/tools/aldente/	database searching and PMF
FindPep	Open source,free	http://www.expasy.ch/tools/findpept.html	database searching and PMF
FindMo	Open source,free	http://www.expasy.ch/tools/findmod	database searching , PM and
MASCO	Commercial	http://www.matrixscience.com	database searching and PMF
ProFoun	Open source,free	http://prowl.rockefeller.edu/prowl-cgi/profound.exe	database searching and PMF
PepNovo	Ope source,free	http://peptide.ucsd.edu/pepnovo.py	DeNov
PEAK	Commercial	http://www.bioinfor.com/peaksonline	DeNov
Lutefis	Open source,free	http://www.hairyfatguy.com/Lutefisk/	DeNov
SEQUES	Commercial	http://fields.scripps.edu/sequest	databas searchin
XTandem	Open source,free	http://www.thegpm.org/tandem	database searching
OMSS	Open source,free	http://pubchem.ncbi.nlm.nih.gov/omssa	database searching
PHENY	Commercial	http://www.phenyx-ms.co	database searching
ProbI	Open source,free	http://www.systemsbiology.org/research/probid/	database searching
Popita	Open source,free	http://www.expasy.org/people/pig/heuristic.html	database searching
Interac	Open source,free	http://www.proteomecenter.org/software.php	database searching
DTAselec	Open source,free	http://fields.scripps.edu/DTASelect/index.html	database searching
Chompe	Open source,free	http://www.ludwig.edu.au/jpsl/jpslhome.html	database searching
ProteinProphet	Open source,free	http://www.proteomecenter.org/software.php	database searching
FindPep	Open source,free	http://www.expasy.ch/tools/findpept.html	database searching
GutenTa	Open source,free	http://fields.scripps.edu/	Denovo and PTM
M -Convolution	fre	http://prospector.ucsf.edu/prospector/mshome.htm	DeNov and
M -Alignment	fre	http://prospector.ucsf.edu/prospector/mshome.htm	DeNov and

Table 2. Summary of the most recognized softwares employed for protein analysis.

For instance one of the most employed algorithm for PMF is Aldente (http://www.expasy.ch). This software allows the protein identification in multi-step way. In the first step the most statistically significant proteins are identified on the basis of accurate peptide m/z combination. In the second one the peptide m/z ion leading to the first identification are not considered and other spectra m/z signal combination are considered in order to identify other proteins. The step is reaped since the identification statistic is good enough in order to identify the protein candidates. In the case b) (MS/MS) the peptides are fragmented using different kind of chemical physical reactions [collision induced dissociation (CID), electron transfer dissociation (ETD), ecc]. The m/z ratio of the peptide fragments is then analyzed in order to obtain peptide structural information. Two approaches are usually employed in order to elaborate the fragmentation spectra: database search and *de novo* sequence. In the case of database search, the peptide MS/MS fragmentation spectra are matched against the theoretical one extracted from public or private repositories. The peptide sequence identification is obtained on the basis of a similarity score among the experimental MS/MS and the theoretical MS/MS spectra. The main limitation of this approach is that only known proteins, reported in the database can be identified. For instance, Thegpm (The Global Proteome Machine; http://www.thegpm.org) is an open source project aims to provide a wide range of tools for proteome identification. In particular X!Tandem software (Muth et al., 2010) is widely employed for database search protein identification. When the protein sequence is not perfectly known, denovo sequence method can be used. In this case, the sequence is obtained directly from the MS/MS spectra avoiding the step of database spectrum search. The obtained sequences are then compared with those contained in the database so to detect homologies. Even in this case a statistical protein identification score is calculated on the basis of the number of homologues fragments obtained for each protein candidate. The software usually employed in the case of database search approach are classified in table 2 together with those employed for *de novo*. An example of software used for de novo porpoises is PepNovo (Frank et al., 2005). It has been presented a novel scoring method for de novo interpretation of peptides from tandem mass spectrometry data. Our scoring method uses a probabilistic network whose structure reflects the chemical and physical rules that govern the peptide fragmentation. The algorithm was tested on ion trap data and achieved results were comparable and in some cases superior to classical database search algorithms. Moreover, different elaborative approaches have been developed in order to increase the sample throughput and statistical accuracy of the identification process (Jacob et al., 2010). Various statistical validation algorithms have been translated into binary programs and are freely distributed on the internet (table 1). Others are not freely available while some have been theoretically described but have not been translated into a freely available or commercial binary program (table 1). It must be stressed that open-source and freely available programs are capable of highly accurate statistical analysis. For example, an interesting free program referred to as ProbID (Zhang et al., 2002) is freely available for evaluation. This program is based on a new probabilistic model and score function that ranks the quality of the match between the peptide. ProbID software has been shown to reach performance levels comparable with industry standard software. A variety of other software, based on heuristic or similar to ProbID Bayesian approach have been developed (Jacob et al., 2010). Some of these software are reported in table 2. It must be stressed that many of these software packages require a web server to operate (e.g., ProbID). This fact introduces some problems related to the difficulty to administrate a server, especially from a security point of view in the case of cracker informatic attacks to a chemstation connected to the internet.

The analysis of intact proteins (top down approach) can be an alternative to bottom up one (Cristoni et. al., 2004). In the first step of data elaboration, the molecular weight of the protein is obtained using dedicated deconvolution algorithm. For instance, Zheng and coworkers have proposed a new algorithm for the deconvolution of ESI mass spectra based on direct assignment of charge to the measured signal at each m/z value in order consequently indirectly to obtain the protein molecular weight (Zheng H, et al. 2003). Another interesting deconvolution approaches is based on the free software named MoWeD (Lu et al., 2011). It can be used to rapidly process LC/ESI/MS data to assign a molecular weight to peptides and proteins. It must be stressed that, the list of found components can also be compared with a user defined list of target molecular weight values making it easy to identify the different proteins present in the analyzed samples. However, when the protein sample mixture is highly complicated, these software tools could fail. This occurs especially if the analysis is performed using low mass accuracy instruments (e.g., IT) and if the chromatographic separation performed before MS analysis is not optimal. Thus, the molecular weight data must be integrated with protein sequence information. In this case, intact proteins ions are analyzed and fragmented by means of high resolution and mass accuracy mass analyzer (e.g.: FTICR, orbitrap, QTOF etc;). The mass spectra obtained are matched directly with the theoretical one present in the database and a statistical score, based on the spectra similarity, is associated with the protein identification. The main advantage of this technology is the ability to investigate intact proteins sequence directly avoiding time consuming digestion steps. On the other hand the majority of algorithm are usually developed for bottom up approach. In fact for different chemical physical reasons, that are not related to this chapter theme, the sensitivity in detecting high molecular weight proteins is definitely lower with respect to that obtained by detecting low molecular weight peptide after protein digestion. An example of algorithm for protein identification, by intact protein ion fragmentation, has been proposed by McLafferty and co-workers (Sze et al., 2002). A free web interface to be used to analyze proteins MS data using the top-down algorithm is available free of charge for academic use. In the top-down MS approach, the multicharged ions of proteins are dissociated and the obtained fragment ions are matched against those predicted from the database protein sequence. This is also a very powerful tool to characterize proteins when complex mixtures are available.

3.3 Data elaboration for functional proteome
Functional proteome (May et al., 2011) is related to both identify differentially expressed proteins among different sample lines and obtain their relative quantitation. For instance, it is possible to compare differentially expressed proteins among control and unhealthy subjects affected by different diseases (Nair et al., 2004). The classical approach (Figure 3) is based on the protein separation by means of 2D-GEL electrophoresis.

The protein are then colored by using specific reagent (e.g. blue coumassie, silver stain etc) and the gel images are obtained by means of a normal or laser fluorescence scanner. Specific software are then employed in order to overlap the images and detect the differentially expressed proteins on the basis of the color intensities. This approach, has strong limitations mainly in terms of elaboration time needed to obtain the match. Nowadays some apparatus have been developed in order to mark, with different label fluorescence reagents, the proteins extracted from different spots. Thus it is possible to run more samples at the same time and detect the proteins of more spots, separately, by means of different fluorescence laser. Distinctive images relative to different gradient of fluorescence are so simultaneously obtained, this results in differentially expressed proteins.

High innovative shut-gun technology based on liquid chromatography coupled to high resolution mass spectrometry, have been recently developed and employed for functional proteomics purposes. In particular, to compare a complex protein mixture of different experimental lines, the obtained peptides after digestion have been analyzed by means of Surface Activated Chemical Ionization (SACI; Cristoni et al. 2007) technologies coupled to high relation and mass accuracy mass analyzer (e.g. Orbitrap, QTOF etc). Very recently *SACI* technology has been applied in seagrass proteomics (Finiguerra et al., 2010). In fact, the increasing sensitivity of this ionization device improves peptides detection thus recovering the limited sea grass genome resources. SACI leads to benefits, in complex plant protein mixture analysis, in terms of quantitative accuracy, precision, and matrix effect reduction, that have been widely demonstrated (Cristoni et al., 2009). As regard peptide analysis, it was observed that, by changing in-source ionization conditions, one could selectively produce both in-source singly and doubly charged species (Cristoni et al., 2007), which are both of interest. This technologic approach yields maximum benefits when data are acquired using a high mass-accuracy and high-resolution mass analyzer that can operate in both full-scan and tandem mass spectrometry (MS/MS) acquisition conditions. The SACI technique strongly increased the number of detectable proteins and of assigned peptides for each protein. For example, with SACI technology application, it was possible to identify a previously identified protein (a heat shock cognate protein), 1000 fold over expressed in deeper plants (-27 m) in comparison with the more shallow plants (-5m), detecting four peptides respect to only two detected by micro-ESI (Finiguerra et al., 2010).

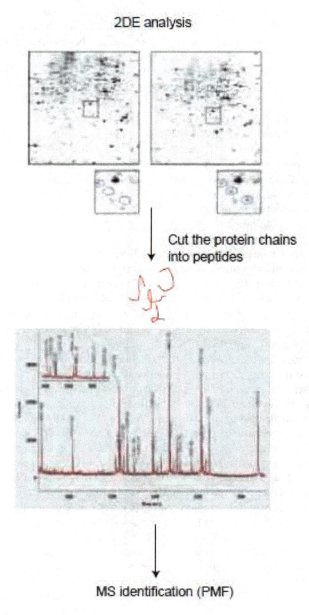

Fig. 3. Classical approach for functional proteome analysis.

The differentially expressed peptides are compared using specific chromatographic alignment software (Sandin et al., 2011). One of the most effective software is named XCMS (Margaria et al., 2008; Kind et al., 2007). The peptide mass fingerprint of the differentially expressed peptides followed by database search and de novo sequencing approach lead to the rapid identifications of differentially expressed proteins.

3.4 Data elaboration for the study of post translational modification

Post translational modification (PTM) detection and quantization is one of the most difficult task in proteomics research. Usually, they are detected through bottom up approaches. For example, considering that phosphorylated peptides do not show a high quality signal intensity, consequently leading to lower sensitivity, dedicated informatics tools have been developed in order to characterize the phosforilation sites on the basis of peptides fragmentation spectra. Different tools developed for this purpose are reported in table 1. All tools detect the modified sites on the basis of fragmentation spectra. Basically, the MS/MS spectra of the theoretical modified peptides are calculated and matched with the experimental one. Even in this case the similarity score is used in order to identified the peptides and the corresponding phosphorylation site.

In the case of PTM the classical PMF and database search approaches cannot be used due to the fact that the modifications and the mutations cause shifts in the MS/MS peaks. Several approaches use an exhaustive search method to identify and characterize the mutated peptides. A virtual database of all modified peptides for a small set of modifications is generated and the peptide MS/MS spectrum is matched against the theoretical spectra of the virtually modified peptides. However, the MS/MS peak shifts result in an increase in the search space and a long elaboration time could be required. To solve this problem some algorithms have been developed (Cristoni S, Bernardi LR. et al. 2004). A good method to detect the mutated peptides is based on the tags approach. For example, the GutenTag software developed by Yates and coworkers use this strategy to identify and characterize the mutated peptides (Tabb DL, et al. 2003). This software infers partial sequences ('tags') directly from the fragment ions present in each spectrum, examines a sequence database to assemble a list of candidate peptides and evaluates the returned peptide sequences against the spectrum to determine which is the best match. The software, written in the Java programming language, runs equally well under Microsoft Windows, Linux and other operating systems. GutenTag is specific to doubly charged peptides. Pevzner and coworkers have also developed some interesting algorithms for the analysis of peptide mutations (Pevzner PA, et al. 2001). In this case, two software packages (MS-CONVOLUTION and MS-ALIGNMENT) that implement the spectra (table 2) convolution and spectral alignment approaches, to identify peptides obtained through enzymatic digestion, have been used to identify and characterize peptides differing by up to two mutations/modifications from the related peptide in the database. This is a two-stage approach to MS/MS database searching. At the first stage, the spectral alignment is used as a filter to identify t, top-scoring peptides in the database, where t is chosen in such a way that it is almost guaranteed that a correct hit is present among the top t list. These top t hits form a small database of candidate peptides subject to further analysis at the second stage. At the verification stage, each of these t peptides can be mutated (as suggested by spectral alignment) and compared against the experimental spectrum. However, the peptide mutation or modification can produce low informative fragmentation behavior (Cristoni et al., 2004), in which case the protein modification identification may fail. It is also possible to use the PMF approach to characterize mutations and modifications (Cristoni et al., 2004). In this case, it is necessary to use a high mass accuracy mass spectrometer since the characterization of a mutation or modification is based on the identification of the accurate m/z ratios of digested peptide. Freeware software to identify protein modifications and mutations using database search and PMF are reported in the Information Resources section. For example, the software FindPept is capable of identifying peptides that result from nonspecific cleavage of proteins from their experimental masses, taking into account artifactual chemical modifications, PTM and protease autolytic cleavage. If autolysis is to be taken into account, an enzyme entry must be specified from the drop-down list of enzymes for which the sequence is known. Furthermore, this is a web application installed on the expasy website and therefore it is not necessary to install and administrate it on a local server. Another field in which different algorithms have been employed is the characterization of disulfide cross-link locations (Cristoni et al., 2004). For instance, some tools available on a public website http://www.expasy.ch were recently developed for this purpose. This software is referred to as Protein Disulfide Linkage Modeler and it permits the rapid analysis of mass spectromic disulfide cross-link mapping experiments. The tool can be used to determine disulfide linkages in proteins that have either been completely or partially digested with enzymes. The masses of all possible disulfide-linked multichain peptide combinations are calculated from the known protein sequence and compared with the experimentally determined masses of disulfide-linked multichain peptides. Thus, this software is based on the fragmentation behavior of the cross-linked peptides obtained by enzymatic digestion. However, several issues may occur despite the fact that this algorithm executes its work very well. The major issue is that proteins containing linker cysteines have domains that are very resistant to proteolysis. Furthermore, the fragmentation of the cross-linked peptide ions may lead to a spectra that is difficult to elaborate even if specific algorithms are used. This is due to the high chemical noise level that is present in the fragmentation spectra of their multicharged ions (Craig et al., 2003).

4. Data management - The advent of semantic technologies and machine learning methods for proteomics

For Systems Biology the integration of multi-level Omics profiles (also across species) is considered as central element. Due to the complexity of each specific Omics technique, specialization of experimental and bioinformatics research groups have become necessary, in turn demanding collaborative efforts for effectively implementing cross-Omics (Wiesinger M, et al. 2011).
In recent years large amounts of information have been accumulated in proteomic, genetic and metabolic databases. Much effort has been dedicated to developing methods that successfully exploit, organize and structure this information. In fact semantic is the study of meaning. In the case of proteomics it can be used in order to find specific relations among proteins and metabolomics, genomics and ionomics networks. For instance the group of Masaneta-Villa and co-workers (Massanet-Vila et al., 2010) has developed a high-throughput software package to retrieve information from publicly available databases, such as the Gene Ontology Annotation (GOA) database and the Human Proteome Resource Database (HPRD) and structure their information. This information is presented to the user as groups of semantically described dense interaction subnetworks that interact with a target protein. Another interesting technology in the semantic field has been proposed by

the group of Mewes HW. and co-workers (Mewes et al., 2011). This group has many years of experience in providing annotated collections of biological data. Selected data sets of high relevance, such as model genomes, are subjected to careful manual curation, while the bulk of high-throughput data is annotated by automatic means. This is, in fact an important point, manual curation is essential for semantic technology purposes. The data mean must be carefully checked before of the insertion in the semantic database otherwise serious meaning error can occurs during the research phase. High-quality reference resources developed in the past and still actively maintained include Saccharomyces cerevisiae, Neurospora crassa and Arabidopsis thaliana genome databases as well as several protein interaction data sets (MPACT, MPPI and CORUM). More recent projects are PhenomiR, the database on microRNA-related phenotypes, and MIPS PlantsDB for integrative and comparative plant genome research. The interlinked resources SIMAP and PEDANT provide homology relationships as well as up-to-date and consistent annotation for 38,000,000 protein sequences. PPLIPS and CCancer are versatile tools for proteomics and functional genomics interfacing to a database of compilations from gene lists extracted from literature. A novel literature-mining tool, EXCERBT, gives access to structured information on classified relations between genes, proteins, phenotypes and diseases extracted from Medline abstracts by semantic analysis.

Another interesting semantic application has been shown by Handcock J. and co-workers (Handcock, et al., 2010). This group has semantically correlate proteomics information to specific clinical diseases. They have produced a database mspecLINE. Given a disease, the tool will display proteins and peptides that may be associated with the disease. It will also display relevant literature from MEDLINE. Furthermore, mspecLINE allows researchers to select proteotypic peptides for specific protein targets in a mass spectrometry assay.

Another interesting semantic technology is based on machine learning and is employed for biomarker discovery purposes (Barla et al., 2008).The search for predictive biomarkers of disease from high-throughput mass spectrometry (MS) data requires a complex analysis path. Preprocessing and machine-learning modules are pipelined, starting from raw spectra, to set up a predictive classifier based on a shortlist of candidate features. As a machine-learning problem, proteomic profiling on MS data needs caution like the microarray case. The risk of over fitting and of selection bias effects is in fact, pervasive.

Summarizing semantic technologies can be useful both to correlate the different omics sciences information and to correlate the single omics (e.g. proteomics) to specific information like clinical disease correlated to differentially expressed proteins between control and unhealthy groups (biomarker discovery).

5. Conclusions

Bioinformatics for proteomics has grown significantly in the recent years. The ability of process an high amount of data together with the high specificity and precision of the new algorithm in the protein identification, characterization and quantization make now possible to obtain an high amount of elaborated data.

The main problem remain the data management of a so high amount of data. Find the correlation among different proteomic data and the other omics sciences (metabolomics, genomics, ionomics) still remain a difficult task. However, database technology together with new semantic statistical algorithm are in evolution powerful tools useful to overcome this problem.

6. Acknowledgment

This is in memory of Anna Malomo-Mazzuca and Remo Cristoni, mother and father of the authors.

7. References

Alagna, F.; D'Agostino, N.; Torchia, L.; Servili, M.; Rao, R.; Pietrella, M.; Giuliano. G.; Chiusano, M.L.; Baldoni, L. & Perrotta, G. (2009). Comparative 454 pyrosequencing of transcripts from two olive genotypes during fruit development. *BMC Genomics,* vol. 10, p.399

Agrawal, G.K. & Rakwa, R. (2005). Rice proteomics: a cornerstone for cereal food crop proteomes. *Mass Spectrom. Review,* vol. 25, pp.1–53

Agrawal, G.K.; Bourguignon, J.; Rolland, N.; Ephritikhine, G.;, Ferro, M.; Jaquinod, M.; Alexiou, K.G.; Chardot, T.; Chakraborty, N.; Jolivet, P.; Doonan, J.H. & Rakwal, R. (2010). Plant organelle proteomics: collaborating for optimal cell function. *Mass Spectrometry Reviews,* 30: n/a. doi: 10.1002/mas.20301

Andacht, T.M. & Winn, R.N.(2006). Incorporating proteomics into aquatic toxicology. *Marine Environmental Research.,* vol. 62, pp.156–186

Apraiz, I.; Mi, J. & Cristobal, S.; Identification of proteomic signatures of exposure to marine pollutants in mussels (*Mytilus edulis*). *Molecular and Cellular Proteomics,* (2006), No (5),pp. 1274–1285

Armengaud, J. (2010). Proteogenomics and systems biology: quest for the ultimate missing parts. *Expert Rev Proteomics,* vol. 7, No.1, pp. 65-77.

Baerenfaller, K.; Grossmann, J.; Grobei, MA.; Hull, R.; Hirsch-Hoffmann, M.; Yalovsky, S.; Zimmermann, P.; Grossniklaus, U.; Gruissem, W. & Baginsky, S. (2008). Genome-scale proteomics reveals *Arabidopsis thaliana* gene models and proteome dynamics. *Science,* vol. 16, pp. 938-941

Baginsky, S. (2009). Plant proteomics: concepts, applications, and novel strategies for data interpretation. *Mass Spectrometry Review,* vol.28, pp.93-120

Baginsky, S.; Hennig, L.; Zimmermann, P. & Gruissem W. (2010). Gene expression analysis, proteomics, and network discovery. *Plant Physiol.* Vol.152, No.(2), pp. 402-10

Barbazuk, W.B.; Emrich, S.J.; Chen, H.D.; Li, L. & Schnable, P.S.(2006). SNP discovery via 454 transcriptome sequencing. *Plant J.* vol. 51, pp. 910-918

Barla, A. Jurman, G. Riccadonna, S. Merler, S. Chierici, M. Furlanello, C. 2008. *Brief Bioinform,* 9, 2, pp. 119-128.

Bellin, D.; Ferrarini, A.; Chimento, A.; Kaiser, O.; Levenkova, N.; Bouffard, P. & Delledonne, M. (2009). Combining next-generation pyrosequencing with microarray for large scale expression analysis in non-model species. *BMC Genomics* vol. 24, pp.55.

Bouck, A.M.Y.; Vision, T. (2007). The molecular ecologist's guide to expressed sequence tags *Molecular Ecology,* vol.16, pp.907–924

Claesson, M.J.; O'Sullivan, O.; Wang, Q.; Nikkilä, J.; Marchesi, J.R.; Smidt, H.; de Vos, W.M.; Ross, R.P. & O'Toole. (2009). Comparative analysis of pyrosequencing and a phylogenetic microarray for exploring microbial community structures in the human distal intestine. *PW.PLoS One, vol.* 20, No. 4, pp.e6669

Chen, X.; Karnovsky, A.; Sans, M.D.; Andrews, P.C. & Williams, J.A. (2010). Molecular characterization of the endoplasmic reticulum: insights from proteomic studies. *Proteomics*, vol. 10, pp. 4040-52. doi: 10.1002/pmic.201000234

Chevalier, F. (2010). Highlights on the capabilities of «Gel-based» proteomics. *Proteome Science*, vol. 8, No.(23), doi:10.1186/1477-5956-8-23

Clarke, S.C. (2005). Pyrosequencing: nucleotide sequencing technology with bacterial genotyping applications. *Expert Rev Mol Diagn.*, vol. 5, No.(6), pp.947-53

Costenoble, R.; Picotti, P.; Reiter, L.; Stallmach, R.; Heinemann, M.; Sauer, U. & Aebersold, R. (2011). Comprehensive quantitative analysis of central carbon and amino-acid metabolism in *Saccharomyces cerevisiae* under multiple conditions by targeted proteomics. *Molecular Systems Biology*, vol. 7, pp. 464

Craft, J.A.; Gilbert, J.A.; Temperton, B.; Dempsey, K.E.; Ashelford, K.; et al. (2010). Pyrosequencing of *Mytilus galloprovincialis* cDNAs: Tissue-Specific Expression Patterns. *PLoS ONE*, vol. 5, No. (1): pp.e8875. doi:10.1371/journal.pone.0008875

Craig, R. Krokhin, O. Wilkins, J. Beavis, RC. 2003. Implementation of an algorithm for modeling disulfide bond patterns using mass spectrometry. *J. Proteome Res*, 2, 6, pp. 657–661.

Cristoni, S.; Zingaro, L.; Canton, C.; Cremonesi, P.; Castiglioni, B.; Morandi, S.; Brasca, M.; Luzzana, M. & Battaglia, C. (2009). Surface-activated chemical ionization and cation exchange chromatography for the analysis of enterotoxin A. *Journal of Mass Spectrometry*, vol. 44,pp. 1482-1488

Cristoni, S.; Rubini, S. & Bernardi, L.R. (2007). Development and applications of surface-activated chemical ionization. *Mass Spectrometry Reviews*, vol. 26, pp. 645-656

Cristoni, S. & Bernardi, L.R. (2004). Bioinformatics in mass spectrometry data analysis for proteomics studies. *Expert Rev Proteomics*, vol. 1, No.4, pp. 469-83.

Cristoni, S. & Bernardi, L,R. (2003). Development of new methodologies for the mass spectrometry study of bioorganic macromolecules. *Mass Spectrom Rev*, vol.22, No.6, pp.369-406.

Dunkley, T.P.J.; Hester, S.; Shadforth, I.P.; Runions, J.; Weimar, T.; Hanton, S.L.; Griffin, J.L.; Bessant, C.; Brandizzi, F.; Hawes, C.; Watson, R.B.; Dupree, P.& Lilley, KS. (2006). Mapping the *Arabidopsis* organelle proteome. *PNAS*, vol. 103, pp. 6518-6523

Doyle, S. (2011). Fungal proteomics: from identification to function. *FEMS Microbiol Lett*. doi: 10.1111/j.1574-6968.2011.02292

Emrich, S.J.; Barbazuk, W.B.; Li, L. & Schnable, P.S. (2007). Gene discovery and annotation using LCM-454 transcriptome sequencing. *Genome Res*, vol.17, pp.69-73

Eveland, A.L.; McCarty, D.R. & Koch, K.E. (2008). Transcript Profiling by 3'-Untranslated Region Sequencing Resolves Expression of Gene Families. *Plant Physiol*, vol. 146, pp. 32-44

Feder, M.E. & Walser, J.C. (2005).The biological limitations of transcriptomics in elucidating stress and stress responses. *Journal of Evolutionary Biology*, vol.18, pp.901–910

Ferro, M.; Brugière, S.; Salvi, D.; Seigneurin-Berny, D.; Court, M.; Moyet, L.; Ramus, C.; Miras, S.; Mellal, M.; Le Gall, S.; Kieffer-Jaquinod, S.; Bruley, C.; Garin, J.; Joyard, J.; Masselon, C. & Rolland, N. (2010). AT_CHLORO: A comprehensive chloroplast proteome database with subplastidial localization and curated information on envelope proteins. *Molecular & Cellular Proteomics*, vol. 9, pp. 1063-1084

Finiguerra, A.; Spadafora, A.; Filadoro, D. & Mazzuca, S. (2010). Surface-activated chemical ionization time-of-flight mass spectrometry and labeling-free approach: two powerful tools for the analysis of complex plant functional proteome profiles. *Rapid Communication in Mass Spectrometry*, vol. 24,pp.1155-1160

Frank A, & Pevzner P. (2005). PepNovo: de novo peptide sequencing via probabilistic network modeling. *Anal Chem*, vol.15, No.77, pp. 964-973.

Fukao, Y.; Hayashi, M. & Nishimura, M. (2002). Proteomic analysis of leaf peroxisomal proteins in greening cotyledons of *Arabidopsis thaliana*. *Plant and Cell Physiology*, vol. 43, pp. 689-696.

Gibson, G. (2003).Microarrays in ecology and evolution: a preview. *Molecular Ecology*, vol. 11, pp. 17-24

Giot, L.; Bader, J.S.; Brouwer, C.; Chaudhuri, A.; & 44 others. (2003). A Protein Interaction Map of *Drosophila melanogaster*. *Science*, vol. 302 , pp.1727–1736

Grennan, A.K. (2008). Plastoglobule proteome. *Plant Physiology*, vol. 147, pp. 443–445

Grimplet, J.; Cramer, G.R.; Dickerson, J.A.; Mathiason, K.; Van Hemert, J.; et al. (2009).VitisNet: "Omics" Integration through Grapevine Molecular Networks. *PLoSONE*, vol. 4,no.(12), pp.e8365. doi:10.1371/journal.pone.0008365

Guo, S.; Zheng, Y.; Joung, J.G.; Liu, S.; Zhang, Z.; Crasta, O.R.; Sobral, B.W.; Xu, Y.; Huang, S. & Fei, Z. (2010) Transcriptome sequencing and comparative analysis of cucumber flowers with different sex types. *BMC Genomics*, vol.11, pp.384

Hamady, M.; Lozupone, C. (2007). Knight R Fast UniFrac: Facilitating high-throughput phylogenetic analyses of microbial communities including analysis of pyrosequencing and PhyloChip data. *ISME J.*, *vol*. 4, No.(1), pp.17-27

Handcock, J. Deutsch, EW. Boyle, J. 2010. mspecLINE: bridging knowledge of human disease with the proteome. *BMC Med. Genomics*, 10, 3, pp. 7.

Haynes, P. A; Roberts, TH. Subcellular shotgun proteomics in plants: Looking beyond the usual suspects. *Proteomics*, , 7, 2963 – 2975

Heazlewood, J.L. & Millar, H. (2005). AMPDB: the *Arabidopsis* Mitochondrial protein Database. *Nucleid Acids Research*. Vol. 33,pp.605–610.

Heazlewood J.L. (2011). The Green proteome: challenges in plant proteomics. Front. Plant Sci., vol. 2, pp6. doi:10.3389/fpls.2011.00006;

Heid, C.A.; Stevens, J.; Livak, K.J. & Williams, PM.(1996). Real time quantitative PCR. *Genome Research*, , 6, 986–994

Ioannidis JP.2010. A roadmap for successful applications of clinical proteomics. *Proteomics Clin Appl*. doi: 10.1002/prca.201000096;

Jacob, RJ. (2010). Bioinformatics for LC-MS/MS-based proteomics. *Methods Mol Biol*, vol. 658, pp. 61-91

Jaquinod, M.; Villiers, F.; Kieffer-Jaquinod, S.; Hugouvieux, V.; Bruley, C.; Garin, J. & Bourguignon, J. (2007). A Proteomics Dissection of *Arabidopsis thaliana* Vacuoles Isolated from Cell Culture. *Molecular and Cellular Proteomics*, 20, vol. 6, 394-412

Jamet, E.; Canut, H.; Boudart, G. & Pont-Lezica, R.F. (2006).Cell wall proteins: a new insight through proteomics *Trends in Plant Science*, vol.11, pp. 33-39

Johnson, S.C. & Browman, H.I. (2007). Introducing genomics, proteomics and metabolomics in marine ecology. *Marine Ecology Progress Series*, vol. 33,pp. 247–248

Joyard, J.; Ferro, M.; Masselon, C.; Seigneurin-Berny, D.; Salvi, D.; Garin, J. & Rolland, N. (2010). Chloroplast proteomics highlights the subcellular compartmentation of lipid metabolism. *Prog Lipid Res*. Vol.49, No. (2), pp.128-58

Joshi H.J.; Hirsch-Hoffmann, M.; Baerenfaller, K.; Gruissem, W.; Baginsky, S.; Schmidt, Robert.; Schulze, W. X.; Sun, Q.; van Wijk K.J.; Egelhofer V.; Wienkoop, S.; Weckwerth, W.; Bruley, C.; Rolland,N.; Toyoda, T.; Nakagami, H.; Jones, A.M.; Briggs, S.P.; Castleden, I.; Tanz, S.K.; A. Millar, H; & Heazlewood. J.L. (2011). MASCP Gator: An Aggregation Portal for theVisualization of Arabidopsis Proteomics Data. *Plant Physiology*, Vol. 155, pp. 259–270

Jovanovic, M.; Reiter, L.; Picotti, P.; Lange, V.; Bogan, E.: Hurschler, B.A.; Blenkiron, C.; Lehrbach, N.J.; Ding, X.C.; Weiss, M.; Schrimpf, S.P.; Miska, E.A.; Großhans, H.; Aebersold, R. & Hengartner, M.O. (2010). A quantitative targeted proteomics approach to validate predicted microRNA targets in C. elegans . *Nature methods*, vol.7, N.(10),pp. 837-845

Kind, T. Tolstikov, V. Fiehn, O. Weiss, RH. 2007. A comprehensive urinary metabolomic approach for identifying kidney cancerr. *Anal Biochem*, vol.363, No.2, pp. 185-195.

Lander, E.S. (2011). Initial impact of the sequencing of the human genome. *Nature*, vol. 470, pp.187–197, doi:10.1038/nature09792

Li, Y.C.; Korol, A.B.; Fahima, T. & Nevo, E. (2004). Microsatellites within genes: structure, function, and evolution. *Molecular Biology and Evolution*, vol. 21, pp.991-1007

Li, X.; Pizarro, A.; Grosser, T. (2009). Elective affinities--bioinformatic analysis of proteomic mass spectrometry data. *Arch Physiol Biochem.*, vol. 115, No(5), pp.311-9

Lopez, J.L. (2007). Applications of proteomics in marine ecology. *Marine Ecology Progress Series*, vol.332, pp. 275–279

Lu, W; Callahan, J.H.; Fry, FS.; Andrzejewski, D.; Musser, SM. & Harrington, P.B. (2011). A discriminant based charge deconvolution analysis pipeline for protein profiling of whole cell extracts using liquid chromatography-electrospray ionization-quadrupole time-of-flight mass spectrometry. *Talanta*, vol. 30, No.84, pp. 1180-1187

Maltman, D.J.; Gadd, S.M.; Simon, W.J. & Slabas, A.R.(2007). Differential proteomic analysis of the endoplasmic reticulum from developing and germinating seeds of castor (*Ricinus communis*) identifies seed protein precursor as significant component of the endoplasmic reticulum. *Proteomics*. Vel. 7, pp.1513–1528

Margaria, T. Kubczak, C. Steffen, B. 2008. Bio-jETI: a service integration, design, and provisioning platform for orchestrated bioinformatics processes. *BMC Bioinformatics*, vol. 9, No. 4, pp. S12.

Marguerat, S.; Brian, T.; Wilhelm, B.T. & Bahler, J. (2008). Next-generation sequencing: applications beyond genomes *Biochemical Society Transactions*, vol. 36, pp.1091-1096.

Maynard Smith, J., & Haigh, J. (1974). The hitch-hiking effect of a favourable gene. *Genetical Research*, vol. 23, pp.23–35

May, C.; Brosseron, F; Chartowski, P.; Schumbrutzki, C.; Schoenebeck, B. & Marcus, K. (2011). Instruments and methods in proteomics. *Methods Mol Bio*, vol. 696, pp. 3-26.

Matthiesen, R. & Jensen, O.N. (2008). Methods Analysis of mass spectrometry data in proteomics. *Mol Biol.* Vol. 453, pp.105-22

Massanet-Vila, R. Gallardo-Chacon, JJ. Caminal, P. Perera, A. 2010. An information theory-based tool for characterizing the interaction environment of a protein. *Conf Proc IEEE Eng Med Biol Soc*, 2010, pp. 5529-5532

Metzker, M.L. (2010). Sequencing technologies-the next generation. *Nature Reviews Genetics*, vol. 11, pp.31-46| doi:10.1038/nrg2626

Mewes, HW. Ruepp, A. Theis, F. Rattei, T. Walter, M. Frishman, D. Suhre, K. Spannagl, M. Mayer, KF. Stümpflen, V. Antonov, A. 2011. MIPS: curated databases and comprehensive secondary data resources in 2010. *Nucleic Acids Res,* 39, pp. 220-224.

Molloy, M.P.; Donohoe, S.; Brzezinski, E.E.; Kilby, G.W.; Stevenson, T.I.; Baker, J.D.; Goodlett, D.R. & Gage, D.A. (2005). Large-scale evaluation of quantitative reproducibility and proteome coverage using acid cleavable isotope coded affinity tag mass spectrometry for proteomic profiling. *Proteomics,* vol. 5, pp. 1204-1208, NCBI-dbEST database [http://www.ncbi.nlm.nih.gov/dbEST]

Muth, T.; Vaudel, M.; Barsnes, H.; Martens, L. & Sickmann, A. (2010). XTandem Parser: an open-source library to parse and analyse X!Tandem MS/MS search results. *Proteomics,* vol.10, No.7, pp. 1522-1524.

Nilsson, T,K. & Johansson, C,A. (2004). A novel method for diagnosis of adult hypolactasia by genotyping of the -13910 C/T polymorphism with Pyrosequencing technology. Scand. *J. Gastroenterol.* Vol. 39, pp.287-290

Nair, KS. Jaleel, A. Asmann, YW. Short, KR. Raghavakaimal, S. (2004). Proteomic research: potential opportunities for clinical and physiological investigators. *Am J Physiol Endocrinol Metab,* vol. 286, No.6, pp. 863-874

Novaes, E.; Drost, D.; Farmerie, W.; Pappas, G.; Grattapaglia, D.; Sederoff, R.R. & Kirst, M. (2008).High-throughput gene and SNP discovery in *Eucalyptus grandis,* an uncharacterized genome. *BMC Genomics* vol. 9, pp.312

Nunn, B.L; & Timperman, T.A. (20079. Marine proteomics. *Marine Ecology Progress Series,* vol. 332, pp. 281-289

Oetjen, K.; Ferber, S.; Dankert, I. & Reusch, T.B.H. (2010). New evidence for habitat-specic selection in Wadden Sea *Zostera marina* populations revealed by genome scanning using SNP and microsatellite markers *Marine Biology* vol. 157, pp. 81–89

Ohlmeier S.; Kastaniotis A. J.; Hiltunen J. K. & Bergmann U. (2004) The yeast mitochondrial proteome - A study of fermentative and respiratory growth. *J Biol Chem.,* vol. 279, pp. 3956-3979

Pevzner, PA. Mulyukov, Z. Dancik, V. Tang CL. 2001. Efficiency of database search for identification of mutated and modified proteins via mass spectrometry. *Genome Res,* 11, 2, pp. 290-299.

Powell, M.J; Sutton, J.N.; Del Castillo, C.E & Timperman, AI. (2005). Marine proteomics: generation of sequence tags for dissolved proteins in seawater using tandem mass spectrometry. *Marine Chemistry,* vol. 95, pp.183-198

Porubleva, L.; VanderVelden, K.; Kothari, S.; Oliver, DJ. & Chitnis, PR. (2001).The proteome of maize leaves: use of gene sequences and expressed sequence tag data foridentification of proteins with peptide mass fingerprinting. *Electrophoresis,* vol. 22, pp. 1724-1738

Plomion, C.; Lalanne, C.; Clavero, S.; Meddour, H.; Kohler, A.; and others. (2006). Mapping the proteome of poplar and application to the discovery of drought-stress responsive proteins. *Proteomics* ,vol.6, pp. 6509–6527

Prokopi, , &, Mayr, M. (2011). Proteomics: a reality-check for putative stem cells. *Circ. Res.,* Vol.108, No. (4),pp.499-511

Repetto, O.; Rogniaux , H.; Firnhaber, C.; Zuber, H.; Küster, H.; Larré, C.; Thompson, R. & Gallardo, K. (2008). Exploring the nuclear proteome of *Medicago truncatula* at the switch towards seed filling. *Plant Journal,* vol. 56, pp. 398

Rossignoll, M.; Peltier, J.B.; Mock, H.P; Matros, A.; Maldonado, A.M. & Jorrín, J.V. (2006). Plant proteome analysis: A 2004–2006 update. *Proteomics*, vol. 6, pp.5529–5548

Roth, U.; Razawi, H.; Hommer, J.; Engelmann, K.; Schwientek, T.; Müller, S.; Baldus, S.E.; Patsos, G.; Corfield, A.P.; Paraskeva, C. & Hisch, F.G. (2010). Differential expression proteomics of human colorectal cancer based on a syngeneic cellular model for the progression of adenoma to carcinoma. Proteomics Clin Appl., vol. 4, no.(8-9),pp.748. doi: 10.1002/prca.201090028

Sá-Correia I, Teixeira MC 2010. 2D electrophoresis-based expression proteomics: a microbiologist's perspective. Expert Rev Proteomics. Dec;7(6):943-53;

Sandin, M. Krogh, M. Hansson, K. Levander, F. (2011) Generic workflow for quality assessment of quantitative label-free LC-MS analysis. *Proteomics*, vol. 11, No. 6, pp. 1114-1124

Satkoski, J.A.; Malhi, R.S.; Kanthaswamy, S.; Tito, R.Y.; Malladi V.S. & Smith, D.G. (2008). Pyrosequencing as a method for SNP identification in the rhesus macaque (*Macaca mulatta*). *BMC Genomics* vol.9, 256doi:10.1186/1471-2164-9-256

Schweitzer, B.; Predki, P. & Snyder, M. (2003). Microarrays to characterize protein interactions on a whole-proteome scale. *Proteomics*, vol.3, pp.2190-2199

Shendure, J.; Porreca, G.J.; Reppas, N.B.; Lin, X.; McCutcheon, J.P.; Rosenbaum, A.M.; Wang, M.D.; Zhang, K.; Mitra, R.D. & Church GM: Accurate multiplex polony sequencing of an evolved bacterial genome. *Science*, vol.309, pp.1728-1732

Stinchcombe, J.R.; & Hoekstra, H.E. (2008). Combining population genomics and quantitative genetics: finding the genes underlying ecologically important traits. *Heredity*, vol.100, pp. 158-170.

Sultan M.; Schulz, M.H.; Richard, H.; Magen, A.; Klingenhoff, A.; Scherf, M.; Seifertet M. al. (2010). A Global View of Gene Activity and Alternative Splicing by Deep Sequencing of the Human Transcriptome *Science*, vol. 321, No. (5891): 956-960. DOI: 10.1126/science.1160342

Sze, SK. Ge, Y. Oh, H. McLafferty, FW. 2002. Top-down mass spectrometry of a 29-kDa protein for characterization of any posttranslational modification to within one residue. *Proc Natl Acad Sci U S A*, 99, 4, pp. 1774-1779.

TAIR Database: The Arabidopsis Information Resource [http://www.arabidopsis.org/] webcite Tair_9_pep_ release 2009 06 19)

Tabb, DL. Saraf, A. Yates, JR. 2003. GutenTag: high-throughput sequence tagging via an empirically derived fragmentation model. *Anal Chem*, vol.75, No. 23, pp. 6415-6421.

Tyers, M. & Mann, M. (2003). From genomics to proteomics. *Nature, vol.* 422, pp. 193–197

Trick, M.; Long, Y.; Meng, J. & Bancroft, I. (2009). Single nucleotide polymorphism (SNP) discovery in the polyploid Brassica napus using Solexa transcriptome sequencing. *Plant Biotechnol J*, vol. 7, pp. 334 -346.

Tsugita, A.; & Kamo, M. (1999). 2-D electrophoresis of plant proteins. *Methods in Molecular Bioogy*, vol.112,pp. 95-97.

Tweedie-Cullen, R.Y. & Mansuy, I.M. (2010). Towards a better understanding of nuclear processes based on proteomics. *Amino Acids*, vol. 39, No. (5), pp.1117-30.

Vasemägi, A. & Primmer, C.R. (2005). Challenges for identifying functionally important genetic variation: the promise of combining complementary research strategies. *Molecular Ecology*, vol. 14,pp.3623–3642

Vasemägi, A.; Nilsson, J. & Primmer, C.R. Expressed Sequence Tag-Linked Microsatellites as a Source of Gene-Associated Polymorphisms for Detecting Signatures of Divergent selection in Atlantic Salmon (*Salmo salar* L.). *Molecular Biology and Evolution*, pp. 1067-1073

Vigouroux, Y.; McMullen, M.; Hittinger, C.T.; Houchins, K.; Schulz, L.; Kresovich, S.; Matsuoka, Y. & Doebley, J. (2002). Identifying genes of agronomic importance in maize by screening microsatellites for evidence of selection during domestication. *Proceedings of the National Academy of Sciences*, USA, vol. 99, pp. 9650–9655

Wang, W.; Wang, Y.; Zhang, Q.; Qi, Y. & Guo, D. (2009). Global characterization of *Artemisia annua* glandular trichome transcriptome using 454 pyrosequencing. *BMC Genomics*, vol. 10, pp.465

Weber, A.P.; Weber, K.L.; Carr, K.; Wilkerson, C. & Ohlrogge, J.B. (2007). Sampling the *Arabidopsis* transcriptome with massively parallel pyrosequencing. *Plant Physiol.*, vol 144, pp. 32-42

Wienkoop, S.; Baginsky, S. & Weckwerth, W.J. (2010). *Arabidopsis thaliana* as a model organism for plant proteome research. *Proteomics*, vol. 73, No.(11), pp.2239-48

Wiesinger, M. Haiduk, M. Behr, M. de Abreu Madeira, HL. Glöckler, G. Perco, P. Lukas, A. 2011. Data and knowledge management in cross-Omics research projects. *Methods Mol Biol*, 719, pp. 97-111

Whitehead, A.; & Crawford, D.L. (2006). Neutral and adaptive variation in gene expression. *Proceedings of the National Academy of Sciences of the United States of America*, vol.103, pp.5425–5430

Wright, J.C.& Hubbard, S.J. (2009). Recent developments in proteome informatics for mass spectrometry analysis. *Comb. Chem. High Throughput Screen.*, vol. 12, No.(2), pp.194-202

Yang, S.; Land, M.L.; Klingeman, D.M.; Pelletier, D.A.; Lu, T.Y.; Martin, S.L.; Guo, H.B. & Smith, J.C.; Brown, S.D. (2010).Paradigm for industrial strain improvement identifies sodium acetate tolerance loci in *Zymomonas mobilis* and *Saccharomyces cerevisiae*. *Proc Natl Acad Sci U S A*, vol.107, No. (23), pp10395-400

Yao, X.; Freas, A.; Ramirez, J.; Demirev, P.A. &Fenselau, C. (2001). Proteolytic ^{18}O labeling for comparative proteomics: model studies with two serotypes of adenovirus. *Analytical Chemistry*, vol. 73, pp.2836–2842

Yauk, C.L. & Lynn Berndt, M.L. (2007). Review of the Literature Examining the Correlation Among DNA Microarray Technologies. *Environmental and Molecular Mutagenesis*, vol. 48,pp. 380-394

Yates, J.R.; Ruse, C.I & Nakorchevsky, A. (2009). Proteomics by mass spectrometry: approaches, advances, and applications. *The Annual Review of Biomedical Engineering*, vol.11, pp. 49-79

Zhang, N.; Aebersold, R. & Schwikowski, B. (2002). ProbID: a probabilistic algorithm to identify peptides through sequence database searching using tandem mass spectral data. *Proteomics*, vol. 2, No. 10, pp.1406-1412

Zheng, H.; Ojha, P.C.; McClean, S.; Black, ND.; Hughes, JG. & Shaw, C. (2003). Heuristic charge assignment for deconvolution of electrospray ionization mass spectra. *Rapid Commun Mass Spectrom*, vol.17, No.5, pp. 429-436.

Protein Progressive MSA Using 2-Opt Method

Gamil Abdel-Azim[1,2], Aboubekeur Hamdi-Cherif[1,3]
Mohamed Ben Othman[1,4] and Z.A. Aboeleneen[5]
[1]College of Computer, Qassim University
[2]College of Computer& Informatics, Canal Suez University
[3]Computer Science Department, Université Ferhat Abbas, Setif (UFAS)
[4]The research Unit of Technologies of Information and Communication (UTIC) /
ESSTT [5]College of Computer& Informatics, Zagazig University
[1]Saudi Arabia
[3]Algeria
[4]Tunisia
[2,5]Egypt

1. Introduction

Multiple sequence alignment (MSA) is a very useful tool in designing experiments for testing and modifying the function of specific proteins, in predicting their functions and structures, and in identifying new members of protein families. MSA of deoxyribonucleic acid (DNA), ribonucleic acid (RNA) and protein remains one of the most common and important tasks in Bioinformatics. Textbooks on the algorithms dedicated to sequence alignment appeared more than a decade ago, e.g. (Durbin et al., 1998). Many critical overviews of the existing MSA have been investigated (Notredame, 2002; Kumar & Filipski, 2007). Finding an optimal MSA of a given set of sequences has been identified as a nondeterministic polynomial-time (NP)-complete problem (Wang & Jiang, 1994). The MSA solution, based on dynamic programming, requires $O((2m)^n)$ time complexity; n being the number of sequences, and m the average sequence length. The memory complexity is $O(m^n)$ (Carrillo & Lipman, 1988; Saitou & Nei, 1987). Therefore, carrying out MSA by dynamic programming becomes practically intractable as the number of sequences increases. The dynamic programming algorithm used for optimal alignment of pairs of sequences can easily be extended to global alignment of three sequences. But for more than three sequences, only a small number of relatively short sequences may be analyzed because of the "curse of dimensionality". Despite the existence of many ready-made and operational systems such as *MBEToolbox* (Cai et al., 2006), *Probalign* (Roshan & Livesay, 2006), *Mulan* (Loots & Ovcharenko, 2007), MSA is always an active area of research (Yue et al., 2009). Approximate methods are constantly investigated for global MSA. One class of these methods is the progressive global alignment. The method starts with an alignment of the most-alike sequences and then builds an alignment by adding more and more closely-alike sequences. Progressive alignment was first formulated in (Hogeweg & Hesper, 1984). Progressive alignment, as implemented in some packages such as ClustalW, for instance, represents one the most popular methodology for MSA. However, in ClustalW, alignment is made by the explicit use of the sequences themselves, which certainly represents a heavy computational burden (Thompson et al., 1994). Building on previous works such as (Azim et al., 2010; 2011), and in order to reduce this computational effort, we represent the similarity between the sequences using a descriptor of the sequences instead of the sequences themselves. The main advantage of using the proposed descriptor resides in its short length, namely 20 for proteins and 4 for DNA, irrespective of the sequence length. Based on this idea, a novel descriptor-based progressive MSA, called DescPA, is formulated, and further improved through a 2-opt method resulting in DescPA2. This novel approach positively impacts the computation time for the MSA, as shown in the results. The chapter is organized as follows. In the next section, the description of protein MSA problem is highlighted. Section 3 briefly presents the DescPA steps as a novel methodology using the Hellinger distance and the computation of the probability density functions (PDF) of sequences. Section 4 reports further enhancements through DescPA2 based on a local search method, namely 2-opt. Section 5 reports the results for both DescPA and DescPA2 performance with respect to ClustalW. Finally, concluding remarks and further research are presented.

2. Proteins MSA problem formulation

2.1 MSA at large
2.1.1 The MSA difficulties
MSA is an interdisciplinary problem. It spans three distinct fields, namely statistics, biology and computer science; each of which encompassing technical difficulties, summarized in the choices of :
- the sequences,
- an objective function (i.e., a comparison model),
- the optimization method for that function.

As a result, properly solving these three problems would require an understanding of all three fields mentioned above, which obviously lies far beyond our reach.

2.1.2 Sequence choice issues
The global of MSA methods assume that we are dealing with a set of homologous sequences i.e., sequences sharing a common ancestor. Furthermore, with the exception of some methods (e.g. Morgenstern et al., 1996), MSA solutions require the sequences to be related over their whole length (or at least most of it). When that condition is not met, one has to rely on the use of local MSA methods such as a sampler (Lawrence et al., 1993), among others.

2.1.3 Objective or cost function issue
The objective or cost function is the mathematical formulation of a purely biological problem that lies in the definition of biological correctness. A mathematical function is used for measuring the biological quality of an alignment. This function is referred to as an objective or cost function since it defines the mathematical objective or cost of the search. Given a perfect function, the mathematically optimal alignment will also be biologically optimal. Unfortunately, this is rarely the case, and while the function defines a mathematical optimum, we rarely have a sound argument that this optimum will also be biologically optimal. As a result, an ideal objective or cost function for all situations does not exist, and every available scheme suffers from major drawbacks. Ideally, a perfect objective or cost

function is to be available for every situation. In practice, this is not the case and the user is always left to make a decision when choosing the method that is most suitable to the problem (Durbin et al., 1998).

2.1.4 Optimization issue
The third main issue associated with MSAs is purely computational. If we assume that we have an adequate set of sequences and a biologically perfect objective function, there still remain the optimization of the objective or cost function. This task is far from being trivial. the computation of a mathematically optimal alignment is too complex a task for an exact method to be used (Wang & Jiang, 1994). Even if we consider a function that consists of the maximization of the number of perfect identities within each column, the problem would still remain intractable for more than three sequences. Consequently, all the current implementations of multiple alignment algorithms are heuristics and that none of them guarantee a full optimization.

2.2 Existing MSA optimization algorithms
Algorithms that construct MSA require a cost function as a criterion for constructing an optimal alignment. There exist three categories of MSA optimization; exact, iterative and progressive (Saitou & Nei, 1987). The exact method suffers from inexact sequence alignment (Wang & Li, 2004). Commonly-used techniques remain the iterative and progressive techniques. Most progressive MSA methods heavily rely on dynamic programming to perform multiple alignments starting with the most closely-related sequences and then progressively adding other related sequences to the initial alignment. These methods have the advantage of being fast, simple as well as reasonably sensitive. Their main drawback is that they can be trapped in local minima that stems from the greedy nature of the algorithm (Thompson et al. 2005). The other major drawback is that any progressive MSA solution cannot be globally optimal, since it is heavily influenced by the initial choice. As a result, any error made at any stage in building the MSA, is propagated and builds up through to the final result. Finally, the performance gets worse when all the sequences in the set are rather distantly-related. Despite these limitations, progressive alignment methods are still efficient enough to be implemented on a large scale for hundreds to thousands of sequences. Hence our contribution to their enhancement.

2.3 Progressive strategy
2.3.1 Basic steps
The existence of several progressive programs and packages has broadened up the aligning techniques. The most popular progressive MSA implementation is represented in the ClustalW family (Higgins & Sharp, 1988; Thompson et al., 1994; 2005). The guide tree in the basic progressive strategy is determined by an efficient clustering method such as neighbor-joining (Saitou & Nei, 1987), or un-weighted average distance (Carrillo & Lipman, 1988).

The progressive strategy, also known as tree method, is one of the most widely used heuristic search for MSA. It combines pairwise alignments beginning with the most similar pair and progressing to the most distantly-related, which finally builds up an MSA solution. The basic progressive alignment strategy follows three steps, depicted in Fig. 1, below.

> // *Basic progressive strategy* //
>
> 1. Compute D, a matrix of distances between all pairs of sequences.
> 2. From D, construct a guide tree T.
> 3. Process the progressive alignment: construct MSA by pairwise alignment of partial alignments (profiles) guided by T.

Fig. 1. Basic progressive strategy

2.3.2 Introductory example

Let $S = \{S_1, S_2, \ldots, S_n\}$ be the input sequences and assume that n is at least 2. Let Σ be the input alphabet that form the sequences. We assume that Σ does not contain the gap character '–'. Any set $S' = \{S'_1, S'_2, \ldots, S'_n\}$ of sequences over the alphabet $\Sigma' = \Sigma \cup \{-\}$, is called an alignment of S if the following two properties satisfied :
1. 1. The strings in S' have the same length.
2. 2. Ignoring gaps, sequences S_i' and S_i are identical.

An alignment can be interpreted as a matrix with n rows and m columns; one row for each S_i and one column for each character in Σ'. Two letters of distinct strings are said to be aligned under S if they are placed into the same column.

For example, Figure 2 shows an alignment for three proteins sequences.

$$AS = \begin{bmatrix} A & R & N & - & D & C & Q & E & G & H & I & L & M & F & - & W & T & W & Y & V \\ - & R & - & N & D & C & Q & E & G & H & I & L & M & F & S & - & T & W & Y & V \\ A & R & N & - & D & C & Q & E & G & H & I & L & M & F & S & - & T & W & Y & V \end{bmatrix}$$

Fig. 2. MSA introductory example for three proteins sequences

3. Descriptor-based progressive MSA (DescPA)

3.1 Basic DescPA
3.1.1 Outline

Within the Clustal-like family, we propose a novel measurement method of the similarity between the sequences, which plays an important role in the building of the guide tree. This measurement is based on the calculation of the probability density function (PDF), also called descriptor or feature vector sequence. The descriptor reduces the dimension of the sequence and yields to a faster calculation of the distance matrix and also to the obtainment of a preliminary distance matrix without pairwise alignment in the first step. For achieving this goal, we use a guide tree based on Hellinger distance. This latter is defined between the descriptors and measures the degree of similarity between the sequences.

3.1.2 DescPA steps

We briefly describe the basic steps of the proposed method, referred to as the descriptor-based progressive MSA (DescPA), outlined in Fig. 3, below.

```
                    // DescPA Steps //
1. Read set of proteins sequences.
2. Estimate the distance matrix between all sequences.
3. Construct the guide tree using distance matrix methods.
4. Apply the progressive alignment methods with guide tree.
5. Output the resulting sequences alignments.
```

Fig. 3. Steps for DescPA

3.1.3 Overall architecture of DescPA

As shown in Figures 3 and 4, the proposed algorithm consists of 3 phases similar to ClustalW. The main difference with ClustalW resides in the way in which the distance matrix is built, here based on Hellinger distance. Each sequence descriptor is described by its probability density function (PDF). The guide tree defines the order in which the sequences are aligned in the next stage. There are several methods for building trees, including distance matrix methods and parsimony methods.

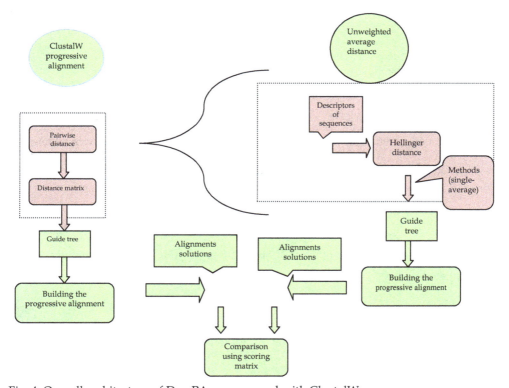

Fig. 4. Overall architecture of DescPA as compared with ClustalW

3.2 Mathematical tools

We need to define some of the basic mathematical tools, necessary for the development of our method. These methods include the Hellinger distance, the PDF calculation, and the scoring matrices.

3.2.1 Hellinger distance

The Hellinger distance is a metric quantity, meaning that it has the properties of non-negativity, identity and symmetry in addition to obeying the triangle inequality. The properties of the Hellinger distance and several related distances are explored in (Donoho & Liu, 1988; Giet & Lubrano, 2008). This concept is used to provide a metric for the distance between two different discrete probability distributions P and Q, as follows:

$$D^2(P,Q) = \frac{1}{2}\sum_{i=1}^{N}(\sqrt{p_i} - \sqrt{q_i})^2 \tag{1}$$

Note that P and Q are described as N-tuples (vectors) of probabilities $(p_1, p_2, ..., p_N)$ and $(q_1, q_2, ..., q_N)$ where p_i and q_i are assumed to be non-negative real numbers with:

$$\sum_{i=1}^{N} p_i = 1; \quad \sum_{i=1}^{N} q_i = 1 \tag{2}$$

3.2.2 Computing the probability density functions (PDFs)

We can compute the Hellinger distance between two variables provided we have explicit knowledge of the probability distributions. Unfortunately, these probabilities are not known in general. Various methods are used to estimate the probability density functions (PDFs) from the observed data. In this paper, we calculate exact probability densities for each proteins sequence. Consider a series x_i and y_i of n simultaneous observations of two random variables X and Y. Since Hellinger distance is computed using discrete probabilities, we proceed as follows:

Let $f_X(i)$ denotes the number of observations i in X. The probabilities p_i are then estimated as:

$$p_i = \frac{f_X(i)}{n} \tag{3}$$

Similarly, let $f_Y(j)$ denote the number of observations of j in Y. The probabilities q_j are then estimated as:

$$q_j = \frac{f_Y(j)}{n} \tag{4}$$

Then the Hellinger distance between X and Y is estimated using Equation (1) above
The descriptor is defined as follows:

$$f : prot \rightarrow [0,1]^n \tag{5}$$

Where *prot* is the set of proteins sequences. The proteins alphabet is given by the 20-character set { A R N D C Q E G H I L K M F P S T W Y V }. The descriptor is calculated for each protein sequence as the PDF of the sequence, obtained as follows:

$$p_i = \frac{N_i}{len(S_i)} \quad (6)$$

where:
$len(S_i)$ is the length of the sequence,
N_i is the number of times character i appears in the sequence.
i belongs to the proteins 20-character alphabet.

3.2.3 Scoring matrices
(i) PAM vs. BLOSUM
Various scoring matrices exist. The main ones are the so-called PAM and BLOSUM (Wheeler, 2003). The most widely used PAM matrix is PAM 250. It has been chosen because it is capable of accurately detecting similarities in the 30% range, that is, when the two proteins are up to 70% different from each other. If the goal is to know the widest possible range of proteins similar to the protein of interest, PAM 250 has been shown to be the most effective. It is also the best to use when the protein is unknown or may be a fragment of a larger protein. Based on an information-theoretic measure called relative entropy it has been shown that the following matrices are equivalent (Henikoff and Henikoff, 1992):
- PAM 250 is equivalent to BLOSUM45.
- PAM 160 is equivalent to BLOSUM62.
- PAM 120 is equivalent to BLOSUM80.

Recall that PAM matrices are the result of computing the probability of one substitution per 100 amino acids, called the PAM1 matrix. Higher PAM matrices are derived by multiplying the PAM1 matrix by itself a defined number of times. Thus, the PAM250 matrix is derived by multiplying the PAM1 matrix against itself 250 times. Biologically, the PAM250 matrix means there have been 2.5 amino acid replacements at each site (Wheeler, 2003).

In the derivation of PAM matrices, sequences that were represented many times were not excluded from the calculation. During the construction of BLOSUM (Blocks Substitution Matrix) matrices, measures were taken to avoid biasing the matrices by removing frequently occurring and highly related sequences. Consequently, as the BLOSUM number decreases (i.e., BLOSUM80, BLOSUM60, BLOSUM50, BLOSUM30...), the ability to detect more distantly related sequences increases in a manner that parallels the effect of increasing the PAM distance (i.e., PAM 40, PAM160, PAM250...), (Altschul, 1991).

(ii) Gonet matrix
In addition to PAM250, we used Gonnet matrix. The Gonnet matrix is a scoring matrix based on alignment of the entire 1991 SwissProt database against itself (Gonnet et al., 1992). A total of 1.7×10^6 matches were used from sequences differing by 6.4 to 100.0 PAM units. This matrix has broad but selective coverage of protein sequences, because SwissProt covers only selected families. This matrix is very useful because of the excellent annotation of proteins included in SwissProt (Wheeler, 2003).

3.2.4 Summarized calculations sub-steps
Fig. 5 below describes the calculations sub-steps undertaken by DescPA.

```
// DescPA calculations sub-steps //
1. Define the descriptor as PDF.
2. Find the PDF for each of two sequences S_i and S_j.
3. Calculate the Hellinger distance between S_i and S_j using Equation (1)
```

Fig. 5. Calculations sub-steps for DescPA

4. Hybridization with 2-opt method

4.1 Local search as improvement methodology

About 93% of the results obtained with basic DescPA compare well with those of ClustalW, as shown in Section 5.1 below, but they are not better. This motivates for the introduction of an enhancement method. A local search method is a good candidate for such an improvement. The resulting improved implementation is referred to as DescPA2. Iterative local search methods rely on algorithms that are able to produce a solution and to refine it, through a series of iterations until no improvement can be made (Wang & Li, 2004), e.g. genetic algorithms as local optimizers (Wang & Lefkowitz, 2005). In our study we propose a local search, that starts from initial solution (i.e. alignment) and repeatedly tries to improve the current solution by local change. If, in the neighborhood of the current alignment a better alignment is found, then it replaces the current one and local search continues. The critical issue in the design of a neighborhood search approach is the choice of the neighborhood structure. In this work, the neighborhood of a solution is depends on the neighborhood $N(\rho)$ of the permutation ρ that is defined by the set of all possible permutations, obtained by exchanging 2 elements. The neighborhood structure $N(PS)$ of the solution is defined as:

$$N(PS) = \bigcup_{(i=1,2,\ldots,n)} N(\rho_i) \tag{6}$$

4.2 The 2-opt method
4.2.1 Outline of the method

The 2-opt method is a combinatorial optimization method originating in the late 1950's in conjunction with the traveling salesman problem (Johnson & McGeoch, 1997). As an adaptation, we define the permutation solution's space corresponding to alignment solution's space. We define the function $\varphi(S,\rho) = S_\rho$ for each sequence S and permutation ρ as follows:

$$S_{\rho(i)} = \begin{cases} S_{\rho'(i)} & i = 1, 2, \ldots, l \\ - & i = l+1, \ldots, m \end{cases} \tag{7}$$

where ρ' is the first sorted elements (sub-permutation) of ρ,
m is the dimension of ρ and l is length of the sequence S. Then by using the definition 7, we can associate permutation solution PS for each alignment solution AS.

4.2.2 Example

Figure 6 illustrates the use of definition 7 with permutation solution *PS*.

Sequence	Length	Permutation	Sub permutation	Sorted permutation
ATCAA	5	(3 5 8 9 1 7 2 4 6)	(3 5 8 9 1)	(1 3 5 8 9)
CGTAGTG	7	(6 7 4 9 1 3 5 2 8)	(6 7 4 9 1 3 5)	(1 3 4 5 6 7 9)
TGATCT	6	(7 6 3 2 1 5 8 9 4)	(7 6 3 2 1 5)	(1 2 3 5 6 7)

Alignment solution	Permutation solution (structure)
$AS = \begin{bmatrix} A-T-C--AA \\ C-GTAGT-G \\ TGA-TCT-- \end{bmatrix}$	$PS = \begin{bmatrix} (3\,5\,8\,9\,1\,7\,2\,4\,6) \\ (6\,7\,4\ 9\,1\,3\,5\,2\,8) \\ (7\,6\,3\ 2\,1\,5\,8\,9\,4) \end{bmatrix}$

Fig. 6. Illustration of the definition of permutation using 3 sequences.

5. Results

5.1 ClustalW *vs.* DescPA results

We compare DescPA with ClustalW using 2 examples. Here, 4 and 9 proteins sequences are used with minimum lengths of 390, and 385 and maximum lengths of 456 and 457, respectively. For both examples, a comparison is made between the results obtained using pairwise (ClustalW) and Hellinger distances (DescPA). We implement the two guide trees using Matlab™ functions as described below.

5.1.1 Guide trees construction

1. **TreePW = seqlinkage(DistancePW,'single',seqs)**, where **seqlinkage** is a Matlab™ function, that implements neighbor-joining algorithm.
2. **DistancePW = seqpdist(seqs,'ScoringMatrix', pam250)**, where **seqs** are the proteins sequences.
3. **TreeHD = seqlinkage (HD,'single',seqs)**, where **HD** is the proposed Hellinger distance matrix.

Figures 7&8 give the comparison between ClustalW (**TreePW**) and DescPA (**TreeHD**) with solution alignment scoring values of the 2 proposed examples over the datasets of BAliBASE 3.0 (Thompson, 2005).

5.1.2 Data set used

The information concerning the data set taken from the database is summarized as follows (Bahr et al., 2001).
Reference 1: Equidistant sequences with 2 different levels of conservation.
Reference 2: Families aligned with a highly divergent "orphan" sequence.
RV11: Reference 1, very divergent sequences (20 identity)
RV12: Reference 1, medium-divergent sequences (20-40 identity).
RV20: Reference 2.
The progressive algorithm is implemented as a Matlab™ function (Version 7.0) called **multialign** which can be used with the following options:

multialign (S, 'terminalGapAdjust', true).

(i) Example 1: Aligning 4 proteins

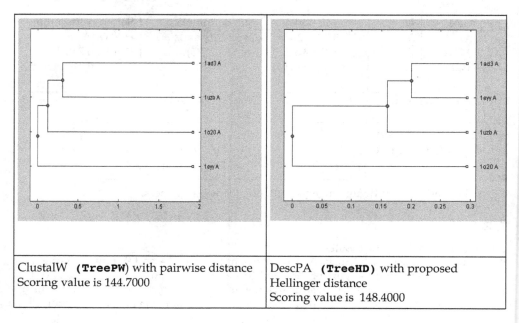

ClustalW **(TreePW)** with pairwise distance Scoring value is 144.7000	DescPA **(TreeHD)** with proposed Hellinger distance Scoring value is 148.4000

Fig. 7. Tree of solutions for ClustalW **(TreePW)** and DescPA **(TreeHD)** for 4 proteins

(ii) Example 2: Aligning 9 proteins

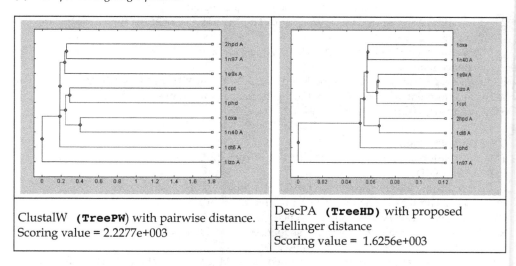

ClustalW **(TreePW)** with pairwise distance. Scoring value = 2.2277e+003	DescPA **(TreeHD)** with proposed Hellinger distance Scoring value = 1.6256e+003

Fig. 8. Tree of solutions for ClustalW **(TreePW)** and DescPA **(TreeHD)** for 9 proteins

5.1.3 DescPA vs. ClustalW Results

Alignments solutions given by the two options (pairwise for ClustalW and Hellinger distance) of the progressive algorithm are implemented in Matlab™ as follows:

(i) Pairwise distance
`SolPW = multialign (seqs, TreePW, 'ScoringMatrix',{ 'pam150 ',' pam200 ',' pam250'})`; where `TreePW` is a guide tree built using pairwise distance.

(ii) Hellinger distance
`SolHD = multialign (seqs, TreeHD, ' ScoringMatrix',{'pam150 ',' pam200 ',' pam250'})`; where `TreeHD` is a guide tree built using the proposed Hellinger distance matrix.

(iii) Results comparison
Figures 9 to 11 show that using the proposed guide tree based on a Hellinger distance gives performance as good as ClustalW in 93% of the cases. To further improve these results, we introduce one iterated local search technique, referred to as 2-opt implemented in Section 5.2.

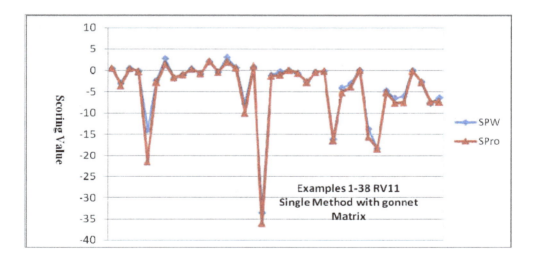

Fig. 9. ClustalW (**SPW**) and DescPA (**Spro**) performance from examples 1-38 (**RV11**)

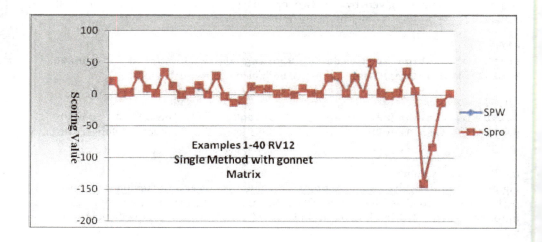

Fig. 10. ClustalW (**SPW**) and DescPA (**Spro**) performance from examples 1-38 (**RV12**)

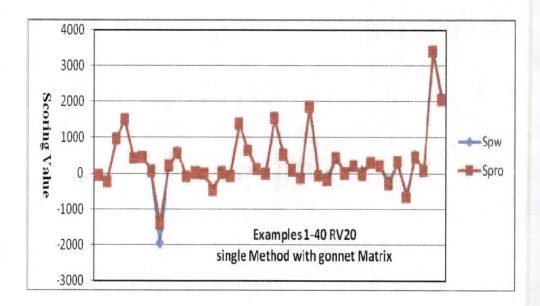

Fig. 11. ClustalW (**SPW**) and DescPA (**Spro**) performance from examples 1-40 (**RV20**)

5.2 DescPA2: Improved results through 2-opt

Figures 12&13 show the improvement on the performance, over different examples of the datasets RV11. There is a clear improvement introduced by the 2-opt algorithm. In Figures 12&13, **SPW** defines the scoring value got using ClustalW, **Spro** gives the scoring value for DescPA and **2-opti** for DescPA2. Despite its simplicity of implementation, the 2-opt algorithm improves the solutions. The final alignments results of DescPA2 are better than those of DescPA and ClustalW.

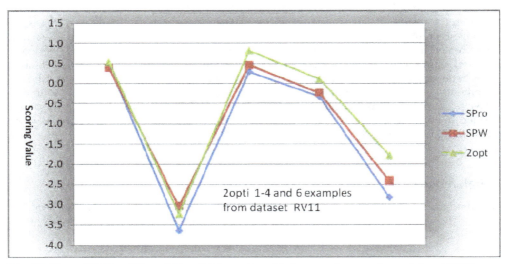

Fig. 12. ClustalW, DescPA and DescPA2 results with 6 examples max from dataset RV11

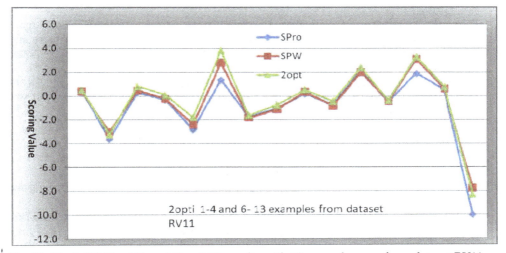

Fig. 13. ClustalW, DescPA and DescPA2 results with 10 examples max from dataset RV11

6. Conclusion

We proposed a modified and hybrid progressive alignment strategy for protein sequence alignment composed of two variants. The first, implemented in DescPA, consists of the modification of the progressive alignment strategy by building a new guide tree based on a Hellinger distance definition. This distance is calculated over a sequences' descriptors; a descriptor being defined for each sequence by its probability density function (PDF). The main feature of this descriptor is its fixed short length (20 for proteins and 4 for DNA) for any sequence length, which mainly impacts positively the computation time for the MSA. The DescPA results of our testing on all the dataset show that the modified progressive alignment strategy is as good as that of ClustalW in 93% of the cases. The second variant, incorporated in DescPA2, is an improvement of the obtained solution using the iterated 2-opt local search. The improvement of the obtained solutions using DescPA2 implementation gives better solutions than DescPA and ClustalW alike - and in all studied cases. As shown, despite its simplicity of implementation, the 2-opt algorithm improves the solutions. However, further improvements are needed. We need, for instance to enhance the actual method to better search through the tree space. For example, we plan to compare DescPA2 with other MSA tools such as hidden Markov models.

7. References

[1] Altschul, S.F. 1991. Amino acid substitution matrices from an information theoretic perspective. *J. Mol. Biol.* (1991), Vol. 219, pp. 555-565, Online ISSN 1460-2059, Print ISSN 1367-4803.

[2] Azim, G.A.; Ben Othman, M. & Abo-Eleneen, Z.A. (2011). Modified progressive strategy for multiple proteins sequence alignment, *International Journal of Computers*. Vol. 5, No.2, (2011), pp. 270-280, ISSN 1998-4308.

[3] Azim, G.A. & Ben Othman, M. (2010). Hybrid iterated local search algorithm for solving multiple sequences alignment problem, *Far East J. of Exp. and Theor. AI*, Vol.5, (2010), pp. 1-17, ISSN 0974-3261.

[4] Bahr, A.; Thompson, J.D.; Thierry, J.C. & Poch O. (2001). BAliBASE (Benchmark Alignment dataBASE): enhancements for repeats, trans-membrane sequences and circular permutations. *Nucleic Acids Res.*, Vol. 29, No. 1, (January 2001), pp. 323-326. Online ISSN 1362-4962, Print ISSN 0305-1048.

[5] Cai, J.J., Smith, D.K., Xia X. & Yuen, K.Y. (2006). *MBEToolbox 2*: an enhanced MATLAB™ toolbox for molecular biology and evolution, *Evol. Bioinf.* Vol. 2, (2006), pp. 189-192, ISSN 1176-9343.

[6] Carrillo, H., & Lipman, D. (1988). The multiple sequence alignment problem in biology. *SIAM J. Appl. Math.* Vol. 48, (1988), pp. 1073-1082, ISSN 0036-1399.

[7] Donoho, D. & Liu, R. (1988). The 'automatic' robustness of minimum distance functionals, *Annals of Stat.*, Vol. 16, (1988), pp. 552-586, ISSN 0090-5364.

[8] Durbin, R. et al. (1998). *Biological Sequence Analysis - Probabilistic Models of Proteins and Nucleic Acids*. Cambridge University Press, ISBN 0521629713, Cambridge, UK.

[9] Giet, L. & Lubrano, M. (2008). A minimum Hellinger distance estimator for stochastic differential equations: an application to statistical inference for continuous time interest rate models, *Comput. Stat. & Data Anal.*, Vol. 52, No. 6, (Feb. 2008), pp. 2945-2965, ISSN: 0167-9473.

[10] Gonnet, G.H.; Cohen, M.A. & Benner, S.A. (1992). Exhaustive matching of the entire protein sequence database. *Science* Vol. 256, (1992), pp. 1443-1445, Online ISSN 1095-9203, Print ISSN 0036-8075.

[11] Higgins D.G., Sharp P.M. (1988). CLUSTAL: a package for performing multiple sequence alignment on a microcomputer. *Gene*, Vol. 73, No. 1, (1988), pp. 237–244, ISSN: 0378-1119.

[12] Henikoff, S. and Henikoff, J.G. 1992. Amino acid substitution matrices from protein blocks. *Proc. Natl. Acad. Sci. U.S.A.Vol.* 89 pp. 10915-10919, Online ISSN 1091-6490, Print ISSN 0027-8424.

[13] Hogeweg, P. & Hesper B. (1984).The alignment of sets of sequences and the construction of phylogenetic trees: an integrated method., *J. Mol. Evol.* Vol. 20, (1984), pp. 175-186, Online ISSN 1432-1432, Print ISSN 0022-2844.

[14] Johnson, D.S. & McGeoch, L.A. (1997). The traveling salesman problem: a case study in local optimization, In Aarts, E.H.L. & Lenstra, J.K. (Eds.) *Local Search in Combinatorial Optimization*, John Wiley and Sons, (1997), ISBN 0471948225, New York, pp. 215-310.

[15] Kumar, S. & Filipski, A. (2007). Multiple sequence alignment: pursuit of homologous DNA positions, *Genome Res.* Vol. 17, (2007), pp. 127-135, ISSN 1088-9051.

[16] Lawrence, C.E.; Altschul S.F.; Boguski M.S.; Liu J.S.; Neuwald A.F. & Wootton J.C. (1993). Detecting subtle sequence signals: a Gibbs sampling strategy for multiple alignment. *Science,* Vol. 262, (1993), pp. 208-214, ISSN 0036-8075.

[17] Loots G. G. & Ovcharenko, I. (2007). *Mulan*: multiple-sequence alignment to predict functional elements in genomic sequences, *Methods Mol. Biol.* 395, (2007), pp. 237-254, ISSN 1064-3745.

[18] Morgenstern, B.; Dress, A.; Wener, T. (1996). Multiple DNA and protein sequence based on segment-to-segment comparison. *Proc. Natl. Acad. Sci. USA* Vol. 93, (1996) pp. 12098-12103, Online ISSN 1091-6490, Print ISSN 0027-8424.

[19] Notredame, C. (2002). Recent progress in multiple sequence alignment: a survey. *Pharmacogenomics*, Vo. 3, No. 1, (2002), pp. 131–144, ISSN 1462-2416.

[20] Roshan, U. & Livesay, D. R. (2006). *Probalign*: multiple sequence alignment using partition function posterior probabilities, *Bioinformatics* Vol. 22, No. 22, (2006), pp. 2715-2721, Online ISSN 1460-2059, Print ISSN 1367-4803.

[21] Saitou N. & M. Nei, (1987). The neighbor-joining method: a new method for reconstructing phylogenetic trees, *Mol. Biol. Evol.*, Vol. 4, No. 4, (1987), pp. 406–425, Online ISSN 1537-1719, Print ISSN 0737-4038.

[22] Thompson, J.D.; Higgins, D.G.; Gibson, T.J. (1994). CLUSTALW: improving the sensitivity of progressive multiple sequence alignment through sequence weighting, position-specific gap penalties and weight matrix choice, *Nucl. Acids Res.* Vol. 22, (1994), pp. 4673–4680, Online ISSN 1362-4962, Print ISSN 0305-1048.

[23] Thompson, J.D.; Koehl, P.; Ripp, R. & Poch, O. (2005). BAliBASE 3.0: latest developments of the multiple sequence alignment benchmark. *Proteins* Vol. 61, (2005) pp.127–136, ISSN 0887-3585.

[24] Wang, Y., & Li, K.-B., (2004). An adaptive and iterative algorithm for refining multiple sequence alignment, *Comput. Biol. Chem.* Vol. 28, (2004), pp. 141–148, ISSN 1476-9271.

[25] Wang, L., & Jiang, T. (1994). On the complexity of multiple sequence alignment. *J. Comput. Biol.* Vol. 1, (1994), pp. 337–348, Online ISSN 1557-8666, Print ISSN 1066-5277.
[26] Wang, C. & Lefkowitz, E.J. (2005). Genomic multiple sequence alignments: refinement using a genetic algorithm, *BMC Bioinformatics*, Vol. 6:200, ISSN 1471-2105.
[27] Wheeler, D. (2003). Selecting the right protein-scoring matrix, *Current Protocols in Bioinformatics* (2003) pp. 351-356, Online ISSN 1934-340X, Print ISSN 1934-3396.
[28] Yue, F., Shi J. & Tang, J. (2009). Simultaneous phylogeny reconstruction and multiple sequence alignment, *BMC Bioinformatics* (2009), Vol. 10 (Suppl. 1):S11, ISSN 1471-2105.

Evolutionary Bioinformatics with a Scientific Computing Environment

James J. Cai
*Texas A&M University,
College Station, Texas
USA*

1. Introduction

Modern scientific research depends on computer technology to organize and analyze large data sets. This is more true for evolutionary bioinformatics — a relatively new discipline that has been developing rapidly as a sub-discipline of bioinformatics. Evolutionary bioinformatics devotes to leveraging the power of nature's experiment of evolution to extract key findings from sequence and experimental data. Recent advances in high-throughput genotyping and sequencing technologies have changed the landscape of data collection. Acquisition of genomic data at the population scale has become increasingly cost-efficient. Genomic data sets are accumulating at an exponential rate and new types of genetic data are emerging. These come with the inherent challenges of new methods of statistical analysis and modeling. Indeed new technologies are producing data at a rate that outpaces our ability to analyze its biological meanings.

Researchers are addressing this challenge by adopting mathematical and statistical software, computer modeling, and other computational and engineering methods. As a result, bioinformatics has become the latest engineering discipline. As computers provide the ability to process the complex models, high-performance computer languages have become a necessity for implementing state-of-the-art algorithms and methods.

This chapter introduces one of such emerging programming languages — Matlab. Examples are provided to demonstrate Matlab-based solutions for preliminary and advanced analyses that are commonly used in molecular evolution and population genetics. The examples relating to molecular evolution focus on the mathematical modeling of sequence evolution; the examples relating to population genetics focus on summary statistics and neutrality tests. Several examples use functions in toolboxes specifically developed for molecular evolution and population genetics — MBEToolbox (Cai, Smith et al. 2005; Cai, Smith et al. 2006) and PGEToolbox (Cai 2008). The source code of some examples is simplified for the publication purpose.

2. Starting Matlab

Matlab is a high-level language and computing environment for high-performance numerical computation and visualization. Matlab integrates matrix computation, numerical analysis, signal processing, and graphics in an easy-to-use environment and simplifies the process of solving technical problems in a variety of disciplines. With Matlab, users access very extensive libraries (i.e., toolboxes) of predefined functions to perform computationally intensive tasks faster than with traditional programming languages such as C, C++, and Fortran. Over the years, Matlab has evolved as a premier program in industrial and educational settings for solving practical engineering and mathematical problems. Researchers in bioinformatics are increasingly relying on Matlab to accelerate scientific discovery and reduce development time.

2.1 Creating & manipulating vectors and matrices

Matlab was designed in the first instance for the purposes of numerical linear algebra. Since its conception, it has acquired many advanced features for the manipulation of vectors and matrices. These features make Matlab an ideal computing language for manipulating genomic data. The basic data type of Matlab is the matrix. Many commonly used genomic data, such as sequences, genotypes, and haplotypes, can be naturally represented as numeric matrices in the computer memory. Therefore, highly efficient basic functions of Matlab can be applied directly to handling many kinds of genomic data. Here is an example of aligned DNA sequences:

```
Seq1    ATCAGGCATCGATGAATCGT
Seq2    ATCGGGCATCGATCAAGCGT
Seq3    ATCGGTCATCTATGAAGGCT
Seq4    ATCGGTCATCGAAGAAGGCG
Seq5    ATCGGTCATCGATCAAGGCG
```

As these sequences are in the same length and are aligned, the alignment can be represented by a Matlab matrix of integers:

```
seq=[1 4 2 1 3 3 2 1 4 2 3 1 4 3 1 1 4 2 3 4
     1 4 2 3 3 3 2 1 4 2 3 1 4 2 1 1 3 2 3 4
     1 4 2 3 3 4 2 1 4 2 4 1 4 3 1 1 3 3 2 4
     1 4 2 3 3 4 2 1 4 2 3 1 1 3 1 1 3 3 2 3
     1 4 2 3 3 4 2 1 4 2 3 1 4 2 1 1 3 3 2 3];
```

The simple mapping converts nucleotide sequences from letter representations (A, C, G, and T) to integer representations (1, 2, 3, and 4). Similarly, genotypic data can be converted into a matrix of integers. The genotypic data below contains nine markers (SNPs) sampled from eight diploid individuals.

```
Idv1    CT GT AG AT AG AG CT AG AG
Idv2    CT GT AG AT AG AG CT AG AG
Idv3    CC GG GG AA AA AA TT GG GG
Idv4    TT TT AA TT GG GG CC AA AA
Idv5    CT GT AG AT AG AG CT AG AG
Idv6    CT GT AG AT AG AG CT AG AG
Idv7    CC GG GG AA AA AA TT GG GG
Idv8    CT GT AG AT AG AG CT AG AG
```

This genotypic data can be converted into the following matrix of integers:

```
geno=[2 4 3 4 1 3 1 4 1 3 1 3 2 4 1 3 1 3
      2 4 3 4 1 3 1 4 1 3 1 3 2 4 1 3 1 3
      2 2 3 3 3 3 1 1 1 1 1 4 4 3 3 3 3
      4 4 4 4 1 1 4 4 3 3 3 3 2 2 1 1 1 1
      2 4 3 4 1 3 1 4 1 3 1 3 2 4 1 3 1 3
      2 4 3 4 1 3 1 4 1 3 1 3 2 4 1 3 1 3
      2 2 3 3 3 3 1 1 1 1 1 4 4 3 3 3 3
      2 4 3 4 1 3 1 4 1 3 1 3 2 4 1 3 1 3];
```

Structures and cell arrays in Matlab provide a way to store dissimilar types of data in the same array. In this example, information about markers, such as the chromosomal position and SNP identification, can be represented in a structure called mark:

```
mark.pos=[38449934,38450800,38455228,38456851,38457117,38457903,...
          38465179,38467522,38469351];

mark.rsid={'rs12516','rs8176318','rs3092988','rs8176297',...
           'rs8176296','rs4793191','rs8176273','rs8176265',...
           'rs3092994'};
```

In the same way, you can represent haplotypes with an integer matrix, hap, and represent makers' information of the haplotype with a mark structure. The difference between sequences of hap and seq is that hap usually contains only sites that are polymorphic and chromosomal positions of these sites are likely to be discontinuous; whereas, seq includes both monoallelic and polymorphic sites, which are continuous in their chromosomal position.

Matlab supports many different data types, including integer and floating-point data, characters and strings, and logical true and false states. By default, all numeric values are stored as double-precision floating point. You can choose to build numeric matrices and arrays as integers or as single-precision. Integer and single-precision arrays offer more memory-efficient storage than double-precision. You can convert any number, or array of numbers, from one numeric data type to another. For example, a double-precision matrix geno can be converted into an unsigned 8-bit integer matrix by using command uint8(geno) without losing any information. The output matrix takes only one-eighth the memory of its double-precision version. The signed or unsigned 8-bit integer, like logical value, requires only 1 byte. They are the smallest data types. Sparse matrices with mostly zero-valued elements, such as adjacency matrices of most biological networks, occupy a fraction of the storage space required for an equivalent full matrix.

2.2 Numerical analysis

Matlab has many functions for numerical data analysis, which makes it a well suited language for numerical computations. Typical uses include problem solving with matrix formulations, general purpose numeric computation, and algorithm prototyping. Using Matlab in numerical computations, users can express the problems and solutions just as they are written mathematically—without traditional programming. As a high-level language, Matlab liberates users from implementing many complicated algorithms and commonly used numerical solutions, and allows users to focus on the "real" problems they want to solve, without understanding the details of routine computational tasks. This section introduces three numerical routines: optimization, interpolation, and integration.

2.2.1 Optimization

Matlab built-in functions and specific toolboxes provide widely used algorithms for standard and large-scale optimization, solving constrained and unconstrained continuous and discrete problems. Users can use these algorithms to find optimal solutions, perform tradeoff analyses, balance multiple design alternatives, and incorporate optimization methods into algorithms and models. Here I use two functions fminbnd and fminsearch to illustrate the general solutions to the problems of constrained linear optimization and unconstrained nonlinear optimization, respectively.

Function fminbnd finds the minimum of a single-variable function, $\min_x f(x)$, within a fixed interval $x_1 < x < x_2$. In Matlab,

```
[x,fval]=fminbnd(@fun,x1,x2);
```

returns scalar x a local minimizer of the scalar valued function, which is described by a function handle @fun, in the interval between x1 and x2. The second returning variable fval is the value of the objective function computed in @fun at the solution x. Function fminsearch finds minimum of unconstrained multivariable function using a derivative-free method. As above, the minimum of a problem is specified by $\min_{\vec{x}} f(\vec{x})$, where $\vec{x}$ is a vector instead of a scalar, and f is a function of several variables. In Matlab this is written as:

```
[x,fval]=fminsearch(@fun,x0);
```

where x0 is a vector of initial values of x. Note that fminsearch can often handle discontinuities particularly if they do not occur near the solution.

Depending on the nature of the problem, you can choose to use fminbnd, fminsearch, or other optimization functions to perform likelihood inference. When doing this, you first need to write a likelihood function that accepts initial parameters as inputs. The likelihood function typically returns a value of the negative log likelihood. Input parameters that produce the minimum of the function are those that give the maximum likelihood for the model. Here is an example showing how to use function fminbnd.

```
options=optimset('fminbnd');
[x,fval]=fminbnd(@likefun,eps,20,options,tree,site,model);
```

where @likefun is a function handle of the following negative log-likelihood function:

```
function [L]=likefun(x,tree,site,model)
rate=x(1);
L=-log(treelike(tree,site,model,rate));
```

This function takes four parameters: the evolutionary rate, rate, (which is what we are going to optimize), a phylogenetic tree, a site of alignment of sequences, and a substitution model. fminbnd returns estimate x(1), which is the optimized rate that gives maximum likelihood fval. The function treelike computes the likelihood of a tree for a given site under the substitution model (see Section 3.4 for details).

2.2.2 Interpolation
Interpolation is one of the classical problems in numerical analysis. Here I show how a one dimensional interpolation problem is formulated and how to use the interpolation technique to determine the recombination fraction between two chromosomal positions. The relationships between physical distance (Mb) and genetic distance (cM) vary considerably at different positions on the chromosome due to the heterogeneity in recombination rate. A recombination map correlates the increment of genetic distance with that of physical distance. The distances between two points in a recombination map are defined in terms of recombination fractions. The incremental step of the physical distance is fixed by the distance between each pair of consecutive makers. Given a set of n makers $[x_k, y_k]$, $1 \leq k \leq n$, with $x_1 < x_2 < ... < x_n$, the goal of interpolation is to find a function $f(x)$ whose graph interpolates the data points, i.e., $f(x_k) = y_k$, for $k = 1, 2,..., n$. The general form of Matlab function interp1 is as follows:

yi=interp1(x,y,xi,method)

where x and y are the vectors holding x-coordinates (i.e., the chromosomal positions) and y-coordinates (i.e., the cumulative recombination rate) of points to be interpolated, respectively. xi is a vector holding points of evaluation, i.e., yi=f(xi) and method is an optional string specifying an interpolation method. Default interpolation method 'linear' produces a piecewise linear interpolant. If xi contains two positions on the chromosome, xi=[pos1,pos2], yi computed will contain two values [rec1,rec2]. The local recombination rate (cM/Mb) can then be calculated as (rec2-rec1)/(pos2-pos1).

2.2.3 Integration
The basic problem considered by numerical integration is to compute an approximate solution to a definite integral $\int_a^b f(x)dx$. If $f(x)$ is a smooth well-behaved function, integrated over a small number of dimensions and the limits of integration are bounded, there are many methods of approximating the integral with arbitrary precision. quad(@fun,a,b) approximates the integral of function @fun from a to b to within an error of 1e-6 using recursive adaptive Simpson quadrature. You can use function trapz to compute an approximation of the integral of Y via the trapezoidal method. To compute the integral with unit spacing, you can use Z=trapz(Y); for spacing other than one, multiply Z by the spacing increment. You can also use Z=trapz(X,Y) to compute the integral of Y with respect to X using trapezoidal integration.

2.3 Data visualization & graphical user interfaces
Matlab adopts powerful visualization techniques to provide excellent means for data visualization. The graphics system of Matlab includes high-level commands for two-dimensional and three-dimensional data visualization, image processing, animation, and presentation graphics. The graphic system of Matlab is also highly flexible as it includes

low-level commands that allow users to fully customize the appearance of graphics. Fig. 1 gives some examples of graphic outputs from data analyses in evolutionary bioinformatics.

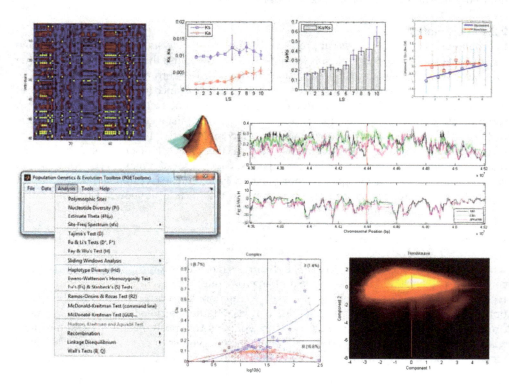

Fig. 1. Examples of graphic outputs and GUIs.

Matlab is also a convenient environment for building graphical user interfaces (GUI). A good GUI can make programs easier to use by providing them with a consistent appearance and intuitive controls. Matlab provides many programmable controls including push buttons, toggle buttons, lists, menus, text boxes, and so forth. A tool called guide, the GUI Development Environment, allows a programmer to select, layout and align the GUI components, edit properties, and implement the behavior of the components. Together with guide, many GUI-related tools make Matlab suitable for application development. PGEGUI and MBEGUI are two menu-driven GUI applications in PGEToolbox and MBEToolbox, respectively.

2.4 Extensibility & scalability

Matlab has an open, component-based, and platform-independent architecture. Scientific applications are difficult to develop from scratch. Through a variety of toolboxes, Matlab offers infrastructure for data analyses, statistical tests, modeling and visualization, and other services. A richer set of general functions for statistics and mathematics allows scientists to manipulate and view data sets with significantly less coding effort. Many special-purpose tool boxes that address specific areas are provided and developers choose only the tools and extensions needed. Thus extensibility is one of the most important features of the Matlab environment. Matlab functions have a high degree of portability, which stems from a complete lack of coupling with the underlying operating system and platform. Matlab application deployment tools enable automatic generation and royalty-free distribution of applications and components. You can distribute your code directly to others to use in their own Matlab sessions, or to people who do not have Matlab.

You can run a Matlab program in parallel. The parallel computing toolbox allows users to offload work from one Matlab session (the client) to other Matlab sessions (the workers). It is possible to use multiple workers to take advantage of the parallel processing on a remote cluster of computers. This is called "remote" parallel processing. It is also possible to do parallel computing with Matlab on a single multicore or multiprocessor machine. This is called "local" parallel computing. Fig. 2 is a screenshot of the task manager showing the CPU usage on a single 8-core PC in local parallel computing with Matlab.

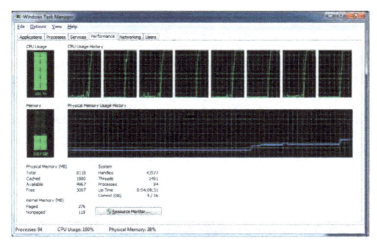

Fig. 2. CPU usage of a single 8-core PC in local parallel computing with Matlab.

With a copy of Matlab that has the parallel computing features, the simplest way of parallelizing a Matlab program is to use the for loops in the program. If a for loop is suitable for parallel execution, this can be indicated simply by replacing the word for by the word parfor. When the Matlab program is run, and if workers have been made available by the matlabpool command, then the work in each parfor loop will be distributed among the workers. Another way of parallelizing a Matlab program is to use a spmd (single program, multiple data) statement. Matlab executes the spmd body denoted by statements on several Matlab workers simultaneously. Inside the body of the spmd statement, each Matlab worker has a unique value of labindex, while numlabs denotes the total number of workers executing the block in parallel. Within the body of the spmd statement, communication functions for parallel jobs (such as labSend and labReceive) can transfer data between the workers. In addition, Matlab is developing new capabilities for the graphics processing unit (GPU) computing with CUDA-enabled NVIDIA devices.

3. Using Matlab in molecular evolution

Molecular evolution focuses on the study of the process of evolution at the scale of DNA, RNA, and proteins. The process is reasonably well modeled by using finite state continuous time Markov chains. In Matlab we obtain compact and elegant solutions in modeling this process.

3.1 Evolutionary distance by counting

Before explicitly modeling the evolution of sequences, let's start with simple counting methods for estimating evolutionary distance between DNA or protein sequences. If two DNA or protein sequences were derived from a common ancestral sequence, then the evolutionary distance refers to the cumulative amount of difference between the two sequences. The simplest measure of the distance between two DNA sequences is the number of nucleotide differences (N) between the two sequences, or the portion of nucleotide differences ($p = N/L$) between the two sequences. In Matlab, the p-distance between two aligned sequences can be computed like this:

```
p=sum(seq(1,:)~=seq(2,:))/size(seq,1);
```

To correct for the hidden changes that have occurred but cannot be directly observed from the comparison of two sequences, the formula for correcting multiple hits of nucleotide substitutions can be applied. The formulae used in these functions are analytical solutions of a variety of Markov substitution models. The simplest model is the JC model (Jukes and Cantor 1969). Analytic solution of the JC model corrects p-distance when $p < 0.75$:

```
d=-(3/4)*log(1-4*p/3);
```

Other commonly used models include Kimura-two-parameter (K2P)(Kimura 1980), Felsenstein (F84)(Felsenstein 1984), and Hasegawa-Kishono-Yano (HKY85)(Hasegawa, Kishino et al. 1985). When the numbers of parameters used to define a model increase with the complicity of the model, we reach a limit where there is no analytical solution for the expression of evolutionary distance. In these cases, we can use the maximum likelihood method, as described in Section 3.3, to estimate the evolutionary distance.

For protein sequences, the simplest measure is the p-distance between two sequences. Assume that the number of amino acid substitutions at a site follows the Poisson distribution; a simple approximate formula for the number of substitutions per site is given by:

```
d=-log(1-p);
```

This is called Poisson correction distance. Given that different amino acid residues of a protein have different levels of functional constraints and the substitution rate varies among the sites, it is suggested that the rate variation can be fitted by the gamma distribution (Nei and Kumar 2000). The gamma distance between two sequences can be computed by:

```
d=a*((1-p)^(-1/a)-1);
```

where a is the shape parameter of the gamma distribution. Several methods have been proposed to estimate a (Yang 1994; Gu and Zhang 1997). The gamma distance with a=2.4

is an approximate of the JTT distance based on the 20×20 amino acid substitution matrix developed by Jones, Taylor et al. (1992). The maximum likelihood estimation of JTT distance is described in Section 3.3.2.

Protein sequences are encoded by strings of codons, each of which is a triplet of nucleotides and specifies an amino acid according to the genetic code. Codon-based distance can be estimated by using the heuristic method developed by Nei and Gojobori (1986). The method has been implemented with an MBEToolbox function called dc_ng86. The function counts the numbers of synonymous and nonsynonymous sites (L_S and L_A) and the numbers of synonymous and nonsynonymous differences (S_S and S_A) by considering all possible evolutionary pathways. The codon-based distance is measured as $K_S = S_S/L_S$ and $K_A = S_A/L_A$ for synonymous and nonsynonymous sites, respectively. Comparison of K_S and K_A provide useful information about natural selection on protein-coding genes: $K_A/K_S = 1$ indicates neutral evolution, $K_A/K_S < 1$ negative selection, and $K_A/K_S > 1$ positive selection.

3.2 Markov models of sequence evolution

Markov models of sequence evolution have been widely used in molecular evolution. A Markov model defines a continuous-time Markov process to describe the change between nucleotides, amino acids, or codons over evolutionary time. Markov models are flexible and parametrically succinct. A typical Markov model is characterized by an instantaneous rate matrix R, which defines the instantaneous relative rates of interchange between sequence states.

R has off-diagonal entries R_{ij} equal to the rates of replacement of i by j: $R_{ij} = r(i \rightarrow j)$, $i \neq j$. The diagonal entries, R_{ii}, are defined by a mathematical requirement that the row sums are all zero, that is, $R_{ii} = \sum_{j \neq i}(-R_{ij})$. The dimension of R depends on the number of statuses of the substitution: 4×4 for nucleotides, 20×20 for amino acids, and 61×61 for codons. We denote Π the vector that contains equilibrium frequencies for 4 nucleotides, 20 amino acids, or 61 sense codons, depending on the model. By multiplying the diagonal matrix of Π, R is transformed into a "frequency-scaled" rate matrix $Q=\text{diag}(\Pi)*R$. Subsequently, we can compute the substitution probability matrix P according to the matrix exponential $P(t) = e^{Qt}$, where $P(t)$ is the matrix of substitution probabilities over an arbitrary time (or branch length) t.

3.3 Model-based evolutionary distance
3.3.1 Nucleic acid substitutions

For a nucleotide substitution probability matrix $P(t)$, $P_{i \rightarrow j}(t)$ is the probability that nucleotide i becomes nucleotide j after time t. An example of divergence of two sequences (each contains only 1 base pair) from a common ancestral sequence is shown in Fig. 3.

Fig. 3. Divergence of two sequences. Sequences 1 (left) and 2 (right) were derived from a common ancestral sequence t years ago. $P_{A \rightarrow C}(t)$ is the probability that nucleotide A becomes C after time t. $P_{A \rightarrow A}(t)$ is the probability that no substitution occurs at the site during time t.

In order to construct the substitution probability matrix P in Matlab, let's first define an instantaneous rate matrix R:

```
>> R=[0,.3,.4,.3;.3,0,.3,.4;.4,.3,0,.3;.3,.4,.3,0]

R =

         0    0.3000    0.4000    0.3000
    0.3000         0    0.3000    0.4000
    0.4000    0.3000         0    0.3000
    0.3000    0.4000    0.3000         0
```

We can use the following command to normalize the rate matrix so that the sum of each column is one:

```
x=sum(R,2); for k=1:4, R(k,:)=R(k,:)./x(k); end
```

This step is unnecessary in this particular example, as original R meets this requirement. Let's assume the equilibrium frequencies of four nucleotides are known (that is, π_A=0.1, π_C=0.2, π_G=0.3, and π_T=0.4).

```
freq=[.1 .2 .3 .4];
```

Here is how to compute and normalize matrix Q:

```
function [Q]=composeQ(R,freq)
PI=diag(freq);
Q=R*PI;
Q=Q+diag(-1*sum(Q,2));
Q=(Q./abs(trace(Q)))*size(Q,1);
```

In Matlab, function EXPM computes the matrix exponential using the Padé approximation. Using this function we can compute substitution probability matrix P for a given time t.

```
P=expm(Q*t);
```

For one site in two aligned sequences, without knowing the ancestral status of the site, we assume one of them is in the ancestral state and the other is in the derived state. If two nucleotides are C and T, and we pick C as the ancestral state, that is, the substitution from C to T, then the probability of substitution $P_{C \to T}(t)$ = P(2,4). In fact, P(2,4) equals to P(4,2), which means the process is reversible. So it does not matter which nucleotide we picked as ancestral one, the result is the same. The total likelihood of the substitution model for the two given sequences is simply the multiplication of substitution probabilities for all sites between the two sequences. In order to estimate the evolutionary distance between two sequences, we try different t-s and compute the likelihood each time until we find the t that gives the maximum value of the total likelihood. This process can be done with optimization functions in Matlab (see Section 2.2.1). The optimized value of t is a surrogate of evolutionary distance between two sequences.

The model of substitution can be specified with two variables R and freq. So we can define the model in a structure:

```
model.R=R;
model.freq=freq;
```

The general time reversible (GTR) model has 8 parameters (5 for rate matrix and 3 for stationary frequency vector). There is no analytical formula to calculate the GTR distance directly. We can invoke the optimization machinery of Matlab to estimate the evolutionary distance and obtain the best-fit values of parameters that define the substitution model.

A convenient method that does not depend on the optimization to compute GTR distance also exists (Rodriguez, Oliver et al. 1990). The first step of this method is to form a matrix F, where F_{ij} denotes the number of sites for which sequence 1 has an i and sequence 2 has a j. The GTR distance between the two sequences is then given by the following formula:

$$d = -tr(\Pi \log(\Pi^{-1}F)),$$

where Π is the diagonal matrix with values of nucleotide equilibrium frequencies on the diagonal, and $tr(X)$ is the trace of matrix X. Here is an example:

```
seq1=[2 3 4 2 3 3 1 4 3 3 3 4 1 3 3 2 4 2 3 2 2 2 1 3 1 3 1 3 3 3];
seq2=[4 2 2 2 3 3 2 4 3 3 2 4 1 2 3 2 4 4 1 4 2 2 1 3 1 2 4 3 1 3];
X=countntchange(seq1,seq2)

X =

     3     0     2     0
     1     4     4     1
     0     0     8     0
     1     3     0     3
```

The formula for computing GTR distance is expressed in Matlab as:

```
F=((sum(sum(X))-trace(X))*R)./4;
F=eye(4)*trace(X)./4+F;
PI=diag(freq);
d=-trace(PI*logm(inv(PI)*F));
```

3.3.2 Amino acid substitutions

For an amino acid substitution probability matrix $P(t)$, $P_{i \to j}(t)$ is the probability that amino acid i becomes amino acid j after time t. In order to compute P, we need to specify the substitution model. As in the case of nucleotides, we need an instantaneous rate matrix model.R and equilibrium frequency model.freq for amino acids. Commonly used R and freq are given by empirical models including Dayhoff, JTT (Jones, Taylor et al. 1992)(Fig. 4), and WAG (Whelan 2008).

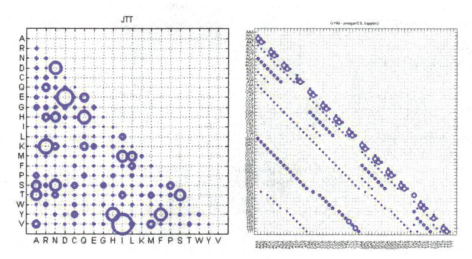

Fig. 4. Visual representations for instantaneous rate matrix **R**. The JTT model of amino acid substitutions (Jones, Taylor et al. 1992) is shown on the left, and the GY94 model of codon substitutions (Goldman and Yang 1994) on the right. The circle size is proportional to the value of the relative rate between pairs of substitutions.

Here I present a function called `seqpairlike` that computes the log likelihood of distance `t` (i.e., branch length, or time) between two protein sequences `seq1` and `seq2` using the model defined with `R` and `freq`. The function `countaachange` is a `countntchange` counterpart for amino acid substitutions.

```
function [lnL]=seqpairlike(t,model,seq1,seq2)
Q=composeQ(model.R,model.freq);
P=expm(Q*t);
X=countaachange(seq1,seq2);
lnL=sum(sum(log(P.^X)));
```

Using the likelihood function, you can adopt an optimization technique to find the optimized `t` as the evolutionary distance between the two sequences.

3.3.3 Codon substitutions

Codon substitutions can be modeled using a Markov process similar to those that are used to describe nucleotide substitutions and amino acid substitutions. The difference is that there are 61 states in the Markov process for codon substitutions as the universal genetic code contains 61 sense codons or nonstop codons. Here I describe a simplified model of Goldman and Yang (1994)(gy94 model). The rate matrix of the model accounts for the transition-transversion rate difference by incorporating the factor κ if the nucleotide change between two codons is a transition, and for unequal synonymous and nonsynonymous substitution rates by incorporating ω if the change is a nonsynonymous substitution. Thus, the rate of relative substitution from codon i to codon j $(i \neq j)$ is:

$$q_{ij} = \begin{cases} 0, & \text{if } i \text{ and } j \text{ differ at two or three codon positions,} \\ \pi_j, & \text{if } i \text{ and } j \text{ differ by a synonymous transversion,} \\ \kappa\pi_j, & \text{if } i \text{ and } j \text{ differ by a synonymous transition,} \\ \omega\pi_j, & \text{if } i \text{ and } j \text{ differ by a nonsynonymous transversion,} \\ \omega\kappa\pi_j, & \text{if } i \text{ and } j \text{ differ by a nonsynonymous transition,} \end{cases}$$

A schematic diagram representing the codon-based rate matrix R with $\omega = 0.5$ and $\kappa = 3.0$ is given in Fig. 4. The function `modelgy94` in MBEToolbox generates the matrix R from given ω and κ:

```
model=modelgy94(omega,kappa);
```

Now let π_j indicate the equilibrium frequency of the codon j. In the GY94 model, $\pi_j = 1/61$, $j = 1, 2, \ldots, 61$.

Here is how we can use GY94 model to estimate dN and dS for two protein-coding sequences. Two sequences are encoded with 61 integers—each represents a sense codon. For example, the following two protein-coding sequences:

```
Seq1    AAA AAC AAG AAT ACA ACC
Seq2    AAT AAC AAG TTA TCA CCC
```

are represented in Matlab with `seq1` and `seq2` like this:

```
seq1=[1 2 3 4 5 6];
seq2=[4 2 3 58 51 22];
```

The codons in original sequences are converted into corresponding indexes in the 61 sense codon list (when the universal codon table is used). This conversion can be done with the function `codonise61` in MBEToolbox: `seq1=codonise61('AATAACAAGTTATCACCC');` You also need a 61×61 mask matrix that contains 1 for every synonymous substitution between codons, and 0 otherwise.

```
% Making a mask matrix, M
T='KNKNTTTTRSRSIIMIQHQHPPPPRRRRLLLLEDEDAAAAGGGGVVVVYYSSSSCWCLFLF';
M=zeros(61);
for i=1:61
for j=i:61
        if i~=j
        if T(i)==T(j) % synonymous change
               M(i,j)=1;
        end
        end
end
end
M=M+M';
```

In the above code, T is the universal code translation table for 61 codons and the corresponding amino acids. Below is the likelihood function that will be used to obtain the three parameters (t, kappa and omega) for the given sequences seq1 and seq2. The input variable x is a vector of [t, kappa, omega].

```
function [lnL]=codonpairlike(x,seq1,seq2)
lnL=inf;
if (any(x<eps)||any(x>999)), return; end
t=x(1); kappa=x(2); omega=x(3);
if (t<eps||t>5), return; end
if (kappa<eps||kappa>999), return; end
if (omega<eps||omega>10), return; end
md=modelgy94(omega,kappa);
R=md.R; freq=md.freq;
Q=composeQ(R,freq);
P=expm(Q*t);
lnL=0;
for k=1:length(seq1)
      s1=seq1(k); s2=seq2(k);
      p=P(s1,s2);
      lnL=lnL+log(p*freq(s1));
end
lnL=-lnL;
```

Given all these, you can now compute the synonymous and nonsynonymous substitution rates per site, d_S and d_N, using maximum likelihood approach:

```
et=0.5; ek=1.5; eo=0.8;    % initial values for t, kappa and omega
options=optimset('fminsearch');
[para,fval]=fminsearch(@codonpairlike,[et,ek,eo],options,seq1,seq2);
lnL=-fval;
t=para(1);
kappa=para(2);
omega=para(3);

% build model using optimized values
md=modelgy94(omega,kappa);
Q=composeQ(md.R,md.freq)./61;

% Calculate pS and pN, assuming omega=optimized omega
pS=sum(sum(Q.*M));
pN=1-pS;

% Calculate pS and pN, assuming omega=1
md0=modelgy94(1,kappa);
Q0=composeQ(md0.R,md0.freq)./61;
pS0=sum(sum(Q0.*M));
pN0=1-pS0;

% Calculate dS and dN
dS=t*pS/(pS0*3);
dN=t*pN/(pN0*3);
```

3.4 Likelihood of a tree

You have learned how to compute the likelihood of substitutions between pairs of sequences. Here I show how to calculate the likelihood of a phylogenic tree given nucleotide sequences. Same technique applies to protein and codon sequences. Imagine you have a tree like the one in Fig. 5. In this example, the four sequences are extremely short, each containing only one nucleotide (i.e., G, A, T, and T). For longer sequences, you can first compute the likelihood for each site independently, and then multiply them together to get the full likelihood for the sequences. The tree describes the evolutionary relationship of the four sequences.

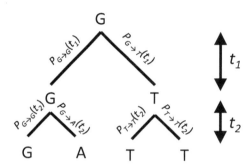

Fig. 5. One path of a tree with 4 external nodes and 3 internal nodes with known states.

Suppose that all internal nodes of the tree are known, which means the ancestral or intermediate states of the site are known. In this case, the likelihood of the tree is:

$$L = P_{G \to G}(t_1) \cdot P_{G \to T}(t_1) \cdot P_{G \to G}(t_2) \cdot P_{G \to A}(t_2) \cdot P_{T \to T}(t_2) \cdot P_{T \to T}(t_2)$$

Thus the likelihood of a phylogenetic tree with known internal nodes at one site can be calculated once the transition probability matrix P is computed as described in Section 3.3.1. In reality, the internal nodes of a tree are unlikely to be known, and the internal nodes can be any of nucleotides. In this case, we need to let every internal node be one of four possible nucleotides each time and compute the likelihood for all possible combinations of nodes. Each distinct combination of nucleotides on all nodes is called a path. Fig. 5 is an instance of one possible path. To get the likelihood of the tree, we multiply all likelihood values (or sum over log likelihood values) that are computed from all possible paths.

Here I use an example to illustrate how to do it using Matlab. Suppose the tree is given in the Newick format:

```
tree='((seq1:0.0586,seq2:0.0586):0.0264,(seq3:0.0586,seq4:0.0586):0.0264):0.043;';
```

The function `parsetree` in MBEToolbox reads through the input tree and extracts the essential information including the topology of the tree, `treetop`, the total number of external nodes, `numnode`, and the branch lengths, `brchlen`.

```
[treetop,numnode,brchlen]=parsetree(tree);
```

The outputs of parsetree are equivalent to the following direct assignment for the three variables:

```
treetop='((1,2),(3,4))';
numnode=4;
brchlen=[0.0586 0.0586 0.0586 0.0586 0.0264 0.0264 0]';
```

Then we prepare an array of transition matrices P. Each transition matrix stacked in P is for one branch. The total number of branches, including both external and internal branches, is 2*numnode-2.

```
n=4;             % number of possible nucleotides
numbrch=2*numnode-2;
P=zeros(numbrch*n,n);
for j=1:numbrch
    P((j-1)*n+1:j*n,:)=expm(Q*brchlen(j));
end
```

In the next step, we use a function called mbelfcreator, which is adapted from Phyllab (Morozov, Sitnikova et al. 2000), to construct an inline function LF. The function mbelfcreator takes two inputs, treetop and numnod, and "synthesizes" the function body of LF. The major operation encoded in the function body is the multiplication of all sub-matrices of the master P matrix. Each sub-matrix is 4×4 in dimension and is pre-computed for the corresponding branch of the tree. The order of serial multiplications is determined by the topology of tree.

```
>>LF=inline(mbelfcreator(treetop,numnode),'P','f','s','n')

LF =

    Inline function:
    LF(P,f,s,n) =
(f*(eye(n)*((P((4*n+1):(5*n),:)*(P((0*n+1):(1*n),s(1)).*P((1*n+1):(2
*n),s(2)))).*(P((5*n+1):(6*n),:)*(P((2*n+1):(3*n),s(3)).*P((3*n+1):(
4*n),s(4)))))))
```

The constructed inline function LF takes four parameters as inputs: P is the stacked matrix, f is the stationary frequency, s is a site of the sequence alignment, and n equals 4 for nucleotide data. With the inline function, we can compute the log likelihood of a site as follows:

```
siteL=log(LF(P,freq,site,n));
```

Finally, we sum over siteL for all sites in the alignment to get the total log likelihood of the tree for the given alignment.
Computing the likelihood of a tree is an essential step from which many further analyses can be derived. These analyses may include branch length optimization, search for best tree, branch- or site-specific evolutionary rate estimation, tests between different substitution models, and so on.

4. Using Matlab in population genetics

Population genetics studies allele frequency distribution and change under the influence of evolutionary processes, such as natural selection, genetic drift, mutation and gene flow. Traditionally, population genetics has been a theory-driven field with little empirical data. Today it has evolved into a data-driven discipline, in which large-scale genomic data sets test the limits of theoretical models and computational analysis methods. Analyses of whole-genome sequence polymorphism data from humans and many model organisms are yielding new insights concerning population history and the genomic prevalence of natural selection.

4.1 Descriptive statistics

Assessing genetic diversity within populations is vital for understanding the nature of evolutionary processes at the molecular level. In aligned sequences, a site that is polymorphic is called a "segregating site". The *number of segregating sites* is usually denoted by S. The expected number of segregating sites $E(S)$ in a sample of size n can be used to estimate population scaled mutation rate $\theta = 4N_e\mu$, where N_e is the diploid effective population size and μ is the mutation rate per site:

$$\theta_W = S\Big/\sum_{i=1}^{n-1}(1/i).$$

In Matlab, this can be written as:

```
[n,L]=size(seq);
S=countsegregatingsites(seq);
theta_w=S/sum(1./[1:n-1]);
```

In the above code, `countsegregatingsites` is a function in PGEToolbox.
Nucleotide diversity, π, is the average number of pairwise nucleotide differences between sequences:

$$\pi = \frac{1}{[n(n-1)/2]}\sum_{i}^{N}\sum_{j<i}^{N}d_{ij},$$

where d_{ij} is the number of nucleotide differences between the ith and jth DNA sequences and n is the sample size. The expected value of π is another estimator of θ, i.e., $\theta_\pi = \pi$.

```
n=size(seq,1);
x=0;
for i=1:n-1
        for j=i+1:n
                d=sum(seq(i,:)~=seq(j,:));
                x=x+d;
        end
end
theta_pi=x/(n*(n-1)/2);
```

Note that, instead of using the straightforward approach that examines all pairs of sequences and counts the nucleotide differences, it is often faster to start by counting the number of copies of each type in the sequence data. Let n_i denote the number of copies of type i, and let $n_{hap} = \sum n_i$. To count the number of copies of the type i, we use the function counthaplotype in PGEToolbox. The general form of the function call is like this:

```
[numHap,sizHap,seqHap]=counthaplotype(hap);
```

where numHap is the total number of distinct sequences or haplotypes, and sizHap is a vector of numbers of each haplotypes. Apparently, sum(sizHap) equals numHap. seqHap is a matrix that contains the distinct haplotype sequences. Using this function, we can calculate nucleotide diversity faster in some circumstances.

```
[nh,ni,sh]=counthaplotype(seq);
x=0;
for i=1:nh-1
for j=i+1:nh
        d=sum(sh(i,:)~=sh(j,:));
        x=x+ni(i)*ni(j)*d;
end
end
theta_pi=x/(n*(n-1)/2);
```

If the sequences are L bases long, it is often useful to normalize θ_S and θ_π by diving them by L. If the genotypic data (geno) is given, the corresponding θ_S and θ_π can be calculated as follows:

```
n=2*size(geno,1);   % n is the sample size (number of chromosomes).
p=snp_maf(geno);    % p is a vector containing MAF of SNPs.
S=numel(p);
theta_w=S/sum(1./(1:n-1));
theta_pi=(n/(n-1))*sum(2.*p.*(1-p));
```

Haplotype diversity (or heterozygosity), *H*, is the probability that two random haplotypes are different. The straightforward approach to calculate *H* is to examine all pairs and count the fraction of the pairs in which the two haplotypes differ from each other. The faster approach starts by counting the number of copies of each haplotype, n_i. Then the haplotype diversity is estimated by

$$H = \frac{1 - \sum_i \left(\frac{n_i}{n_{hap}}\right)^2}{1 - 1/n_{hap}}.$$

Using the function counthaplotype, we can get the number of copies of each haplotype and then compute *H* as follows:

```
[nh,ni]=counthaplotype(hap);
h=(1-sum((ni./nh).^2))./(1-1./nh);
```

Site frequency spectrum (SFS) is a histogram whose ith entry is the number of polymorphic sites at which the mutant allele is present in i copies within the sample. Here, i ranges from 1 to n-1. When it is impossible to tell which allele is the mutant and which is the ancestral one, we combine the entries for i and n-i to make a folded SFS. *Mismatch distribution* is a histogram whose ith entry is the number of pairs of sequences that differ by i sites. Here, i ranges from 0 through the maximal difference between pairs in the sample. Two functions in PGEToolbox, `sfs` and `mismch`, can be used to calculate SFS and mismatch distribution, respectively.

4.2 Neutrality tests

The standard models of population genetics, such as the Wright–Fisher model and related ones, constitute null models. Population geneticists have used these models to develop theory, and then applied the theory to test the goodness-of-fit of the standard model on a given data set. Using summary statistics, they can reject the standard model and take into account other factors, such as selection or demographic history, to build alternative hypotheses. These tests that compute the goodness-of-fit of the standard model have been referred to as "neutrality tests", and have been widely used to detect genes, or genomic regions targeted by natural selection. An important family of neutrality tests is based on summary statistics derived from the SFS. The classical tests in this family include Tajima's D test (Tajima 1989), Fu and Li's tests (Fu and Li 1993), and Fay and Wu's H test (Fay and Wu 2000), which have been widely used to detect signatures of positive selection on genetic variation in a population.

Under evolution by genetic drift (i.e., neutral evolution), different estimators of θ, such as, θ_W and θ_π, are unbiased estimators of the true value of θ: $E(\hat{\theta}_W) = E(\hat{\theta}_\pi) = \theta$. Therefore, the difference between θ_W and θ_π can be used to infer non-neutral evolution. Using this assumption, Tajima's D test examines the deviation from neutral expectation (Tajima 1989). The statistic D is defined by the equation:

$$D = (\theta_\pi - \theta_W)/\sqrt{V(\theta_\pi - \theta_W)},$$

where $V(d)$ is an estimator of the variance of d. The value of D is 0 for selectively neutral mutations in a constant population infinite sites model. A negative value of D indicates either purifying selection or population expansion (Tajima 1989).

```
% n is the sample size; S is the number of segregating sites
% theta_w and theta_pi have been calculated

nx=1:(n-1);
a1=sum(1./nx);
a2=sum(1./nx.^2);
b1=(n+1)/(3*(n-1));
b2=2*(n*n+n+3)/(9*n*(n-1));
c1=b1-1/a1;
c2=b2-(n+2)/(a1*n)+a2/(a1^2);
e1=c1/a1;
e2=c2/(a1^2+a2);
tajima_d=(theta_pi-theta_w)/sqrt(e1*S+e2*S*(S-1));
```

The other SFS-based neutrality tests, like Fu and Li's tests (Fu and Li 1993) and Fay and Wu's H test (Fay and Wu 2000), share a common structure with Tajima's D test. Many other neutrality tests exhibit important diversity. For example, R2 tests try to capture specific tree deformations (Ramos-Onsins and Rozas 2002), and the haplotype tests use the distribution of haplotypes (Fu 1997; Depaulis and Veuille 1998).

4.3 Long-range haplotype tests

When a beneficial mutation arises and rapidly increases in frequency in the process leading to fixation, chromosomes harbouring the beneficial mutation experience less recombination events. This results in conservation of the original haplotype. Several so called long-range haplotype (LRH) tests have been developed to detect long haplotypes at unusually high frequencies in genomic regions, which have undergone recent positive selection.

The test based on the extended haplotype homozygosity (EHH) developed by Sabeti et al. (2002) is one of the earliest LRH tests. EHH is defined as the probability that two randomly chosen chromosomes carrying an allele (or a haplotype) at the core marker (or region) are identical at all the markers in the extended region. EHH between two markers, s and t, is defined as the probability that two randomly chosen chromosomes are homozygous at all markers between s and t, inclusively. Explicitly, if N chromosomes in a sample form G homozygous groups, with each group i having n_i elements, EHH is defined as:

$$EHH = \frac{\sum_{i=1}^{G}\binom{n_i}{2}}{\binom{N}{2}}.$$

Equivalently, EHH can be calculated in a convenient form as the statistic *haplotype homozygosity*:

$$HH = \left(\sum p_i^2 - 1/n\right)\big/(1 - 1/n),$$

where p_i is the frequency of haplotype i and n is the sample size. For a core marker, EHH is calculated as HH in a stepwise manner. The EHH is computed with respect to a distinct allele of a core maker or a distinct formation of a core region. In Fig. 6, for example, we focus on allele A of the core maker (a diallelic SNP) at the position x. Variable hap contains A-carrying haplotypes of size n×m.

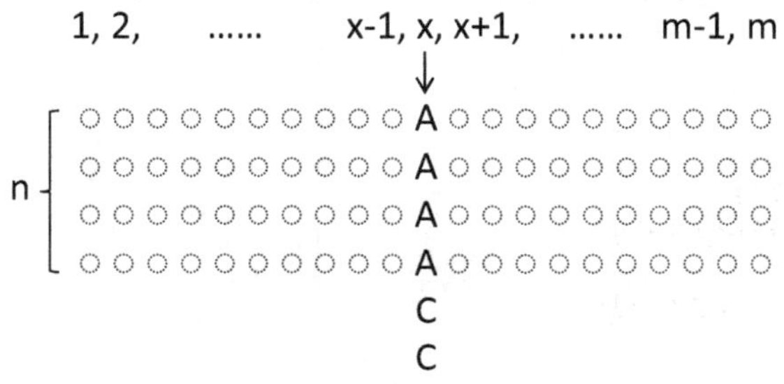

```
            1, 2,    ......   x-1, x, x+1,  ......  m-1, m
                                    ↓
         ⎡ ○ ○ ○ ○ ○ ○ ○ ○ ○ ○ A ○ ○ ○ ○ ○ ○ ○ ○ ○
     n ⎨   ○ ○ ○ ○ ○ ○ ○ ○ ○ ○ A ○ ○ ○ ○ ○ ○ ○ ○ ○
         ⎪ ○ ○ ○ ○ ○ ○ ○ ○ ○ ○ A ○ ○ ○ ○ ○ ○ ○ ○ ○
         ⎣ ○ ○ ○ ○ ○ ○ ○ ○ ○ ○ A ○ ○ ○ ○ ○ ○ ○ ○ ○
                                    C
                                    C
```

Fig. 6. Calculation of EHH for n haplotypes carrying allele A at the focal position x.

The EHH values, ehh1, around x in respect to the allele A, can be computed as follows:

```
ehh1=ones(1,m);
for i=1:x-1
    [n,ni]=counthaplotype(hap(:, i:x-1));
    p=ni./n;
    ehh1(i)=(sum(p.^2)-1/n)/(1-1/n);
end
for j=x+1:m
    [n,ni]=counthaplotype(hap(:, x+1:end));
    p=ni./n;
    ehh1(j)=(sum(p.^2)-1/n)/(1-1/n);
end
```

Similarly, the EHH around x with respect to the allele C, ehh2, can be computed using the same machinery. Both ehh1 and ehh2 are calculated for all markers around the core maker. Fig. 7 shows the EHH curves for two alleles C and T in the core SNP. The EHH values for the markers decrease as the distance from the core marker increases.

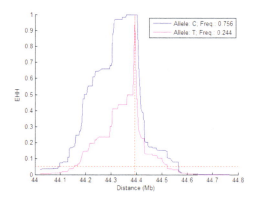

Fig. 7. EHH decay as a function of the distance between a test marker and the core marker. Vertical dash line indicates the location of the core marker. Horizontal dash line indicates the cut-off=0.05 for computing EHH integral.

The integrated EHH (iHH) is the integral of the observed decay of EHH away from the core marker. iHH is obtained by integrating the area under the EHH decay curve until EHH reaches a small value (such as 0.05). Once we obtain ehh1 and ehh2 values for the two alleles, we can integrate EHH values with respect to the genetic or physical distance between the core marker and other markers, with the result defined as iHH$_1$ and iHH$_2$. The statistic $ln(iHH_1/iHH_2)$ is called the integrated haplotype score (iHS), which is a measure of the amount of EHH at a given maker along one allele relative to the other allele. The iHS can be standardized (mean 0, variance 1) empirically to the distribution of the observed iHS scores over a range of SNPs with similar allele frequencies. The measure has been used to detect partial selective sweeps in human populations (Voight, Kudaravalli et al. 2006).

In Matlab, we invoke the function trapz(pos,ehh) to compute the integral of EHH, ehh, with respect to markers' position, pos, using trapezoidal integration. The position is in units of either physical distance (Mb) or genetic distance (cM). The unstandardized integrated haplotype score (iHS) can be computed as the log ratio between the two iHHs:

```
ihh1=trapz(pos,ehh1);
ihh2=trapz(pos,ehh2);
ihs=log(ihh1/ihh2);
```

The cross population EHH (XP-EHH) has been used to detect selected alleles that have risen to near fixation in one but not all populations (Sabeti, Varilly et al. 2007). The statistic XP-EHH uses the same formula as iHS, that is, $ln(iHH_1/iHH_2)$. The difference is that iHH_1 and iHH_2 are computed for the same allele in two different populations. An unusually positive value suggests positive selection in population 1, while a negative value suggests the positive selection in population 2.

4.4 Population differentiation

Genomic regions that show extraordinary levels of genetic population differentiation may be driven by selection (Lewontin 1974). When a genomic region shows unusually high or low levels of genetic population differentiation compared with other regions, this may then be interpreted as evidence for positive selection (Lewontin and Krakauer 1973; Akey, Zhang et al. 2002). The level of genetic differentiation is quantified with F_{ST}, which was introduced by Wright (Wright 1931) measuring the effect of structure on the genetics of a population. There are several definitions of F_{ST} in the literature; the simple concept is $F_{ST} = (H_T - H_S)/H_T$, where H_T is the heterozygosity of the total population and H_S is the average heterozygosity across subpopulations.

Suppose you know the frequencies, p1 and p2, of an allele in two populations. The sample sizes in two populations are n1 and n2. Wright's F_{ST} can be computed as follows:

```
pv=[p1 p2];
nv=[n1 n2];
x=(nv.*(nv-1)/2);
Hs=sum(x.*2.*(nv./(nv-1)).*pv.*(1-pv))./sum(x);
Ht=sum(2.*(n./(n-1)).*p_hat.*(1-p_hat));
Fst=1-Hs./Ht;
```

Below is a function that calculates an unbiased estimator of F_{ST}, which corrects for the error associated with incomplete sampling of a population (Weir and Cockerham 1984; Weir 1996).

```
function [f]=fst_weir(n1,n2,p1,p2)
n=n1+n2;
nc=(1/(s-1))*((n1+n2)-(n1.^2+n2.^2)./(n1+n2));
p_hat=(n1./n).*p1+(n2./n).*p2;
s=2; % number of subpopulations
MSP=(1/(s-1))*((n1.*(p1-p_hat).^2 + n2.*(p2-p_hat).^2));
MSG=(1./sum([n1-1, n2-1])).*(n1.*p1.*(1-p1)+n2.*p2.*(1-p2));
Fst=(MSP-MSG)./(MSP+(nc-1).*MSG);
```

NC is the variance-corrected average sample size, p_hat is the weighted average allele frequency across subpopulations, MSG is the mean square error within populations, and MSP is the mean square error between populations.

5. Conclusion

Matlab, as a powerful scientific computing environment, should have many potential applications in evolutionary bioinformatics. An important goal of evolutionary bioinformatics is to understand how natural selection shapes patterns of genetic variation within and between species. Recent technology advances have transformed molecular evolution and population genetics into more data-driven disciplines. While the biological data sets are becoming increasingly large and complex, we hope that the programming undertakings that are necessary to deal with these data sets remain manageable. A high-level programming language like Matlab guarantees that the code complexity only increases linearly with the complexity of the problem that is being solved.

Matlab is an ideal language to develop novel software packages that are of immediate interest to quantitative researchers in evolutionary bioinformatics. Such a software system is needed to provide accurate and efficient statistical analyses with a higher degree of usability, which is more difficult to achieve using traditional programming languages. Limited functionality and inflexible architecture of existing software packages and applications often hinder their usability and extendibility. Matlab can facilitate the design and implementation of novel software systems, capable of conquering many limitations of the conventional ones, supporting new data types and large volumes of data from population-scale sequencing studies in the genomic era.

6. Acknowledgment

The work was partially supported by a grant from the Gray Lady Foundation. I thank Tomasz Koralewski and Amanda Hulse for their help in the manuscript preparation. MBEToolbox and PGEToolbox are available at http://www.bioinformatics.org/mbetoolbox/ and http://www.bioinformatics.org/pgetoolbox/, respectively.

7. References

Akey, J. M., G. Zhang, et al. (2002). Interrogating a high-density SNP map for signatures of natural selection. *Genome Res* 12(12): 1805-1814.

Cai, J. J. (2008). PGEToolbox: A Matlab toolbox for population genetics and evolution. *J Hered* 99(4): 438-440.

Cai, J. J., D. K. Smith, et al. (2005). MBEToolbox: a MATLAB toolbox for sequence data analysis in molecular biology and evolution. *BMC Bioinformatics* 6: 64.

Cai, J. J., D. K. Smith, et al. (2006). MBEToolbox 2.0: an enhanced version of a MATLAB toolbox for molecular biology and evolution. *Evol Bioinform Online* 2: 179-182.

Depaulis, F. & M. Veuille (1998). Neutrality tests based on the distribution of haplotypes under an infinite-site model. *Mol Biol Evol* 15(12): 1788-1790.

Fay, J. C. & C. I. Wu (2000). Hitchhiking under positive Darwinian selection. *Genetics* 155(3): 1405-1413.

Felsenstein, J. (1984). Distance Methods for Inferring Phylogenies: A Justification. *Evolution* 38(1): 16-24.

Fu, Y. X. (1997). Statistical tests of neutrality of mutations against population growth, hitchhiking and background selection. *Genetics* 147(2): 915-925.

Fu, Y. X. & W. H. Li (1993). Statistical tests of neutrality of mutations. *Genetics* 133(3): 693-709.

Goldman, N. & Z. Yang (1994). A codon-based model of nucleotide substitution for protein-coding DNA sequences. *Mol Biol Evol* 11(5): 725-736.

Gu, X. & J. Zhang (1997). A simple method for estimating the parameter of substitution rate variation among sites. *Mol Biol Evol* 14(11): 1106-1113.

Hasegawa, M., H. Kishino, et al. (1985). Dating of the human-ape splitting by a molecular clock of mitochondrial DNA. *J Mol Evol* 22(2): 160-174.

Jones, D. T., W. R. Taylor, et al. (1992). The rapid generation of mutation data matrices from protein sequences. *Comput Appl Biosci* 8(3): 275-282.

Jukes, T. H. & C. Cantor (1969). Evolution of protein molecules. *Mammalian Protein Metabolism*. H. N. Munro. New York, Academic Press: 21-132.

Kimura, M. (1980). A simple method for estimating evolutionary rates of base substitutions through comparative studies of nucleotide sequences. *J Mol Evol* 16(2): 111-120.

Lewontin, R. C. (1974). *The genetic basis of evolutionary change*. New York,, Columbia University Press.

Lewontin, R. C. & J. Krakauer (1973). Distribution of gene frequency as a test of the theory of the selective neutrality of polymorphisms. *Genetics* 74(1): 175-195.

Morozov, P., T. Sitnikova, et al. (2000). A new method for characterizing replacement rate variation in molecular sequences. Application of the Fourier and wavelet models to Drosophila and mammalian proteins. *Genetics* 154(1): 381-395.

Nei, M. & T. Gojobori (1986). Simple methods for estimating the numbers of synonymous and nonsynonymous nucleotide substitutions. *Mol Biol Evol* 3(5): 418-426.

Nei, M. & S. Kumar (2000). *Molecular evolution and phylogenetics*. Oxford ; New York, Oxford University Press.

Ramos-Onsins, S. E. & J. Rozas (2002). Statistical properties of new neutrality tests against population growth. *Mol Biol Evol* 19(12): 2092-2100.

Rodriguez, F., J. L. Oliver, et al. (1990). The general stochastic model of nucleotide substitution. *J Theor Biol* 142(4): 485-501.

Sabeti, P. C., D. E. Reich, et al. (2002). Detecting recent positive selection in the human genome from haplotype structure. *Nature* 419(6909): 832-837.

Sabeti, P. C., P. Varilly, et al. (2007). Genome-wide detection and characterization of positive selection in human populations. *Nature* 449(7164): 913-918.

Tajima, F. (1989). Statistical method for testing the neutral mutation hypothesis by DNA polymorphism. *Genetics* 123(3): 585-595.

Voight, B. F., S. Kudaravalli, et al. (2006). A map of recent positive selection in the human genome. *PLoS Biol* 4(3): e72.

Weir, B. S. (1996). *Genetic data analysis II : methods for discrete population genetic data*. Sunderland, Mass., Sinauer Associates.

Weir, B. S. & C. C. Cockerham (1984). Estimating F-statistics for the analysis of population structure. *Evolution* 38: 1358-1370.

Whelan, S. (2008). Spatial and temporal heterogeneity in nucleotide sequence evolution. *Mol Biol Evol* 25(8): 1683-1694.

Wright, S. (1931). The genetical structure of populations. *ann Eugenics* 15: 323-354.

Yang, Z. (1994). Maximum likelihood phylogenetic estimation from DNA sequences with variable rates over sites: approximate methods. *J Mol Evol* 39(3): 306-314.

Parallel Processing of Complex Biomolecular Information: Combining Experimental and Computational Approaches

Jestin Jean-Luc and Lafaye Pierre
Institut Pasteur
France

1. Introduction

While protein functions such as binding or catalysis remain very difficult to predict computationally from primary sequences, approaches which involve the parallel processing of diverse proteins are remarkably powerful for the isolation of rare proteins with functions of interest.

Stated using a Darwinian vocabulary, a repertoire of proteins can be submitted to selection according to a function of interest for isolation of the rare fittest proteins. Parallel processing strategies rely mainly on the design of *in vitro* selections of proteins. To ensure that complex molecular information can be extracted after selection from protein populations, several types of links between the genotype and the phenotype have been designed for the parallel processing of proteins: they include the display of nascent proteins on the surface of the ribosome bound to mRNA, the display of proteins as fusions with bacteriophage coat proteins and the fusion of proteins to membrane proteins expressed on the surface of yeast cells. In the first two display strategies, covalent and non covalent bonds define chemical links between the genotype and the protein, while in the last case compartmentation by a membrane provides the link between the protein and the corresponding gene.

While parallel processing strategies allow the analysis of up to 10^{14} proteins, serial processing is convenient for the analysis of tens to thousands of proteins, with the exceptions of millions of proteins in the specific case where fluorescent sorting can be adapted experimentally.

In this review, the power of parallel processing strategies for the identification of proteins of interest will be underlined. It is useful to combine them with serial processing approaches such as activity screening and the computational alignment of multiple sequences. These molecular information processing (MIP) strategies yield sequence-activity relationships for proteins, whether they are binders or catalysts (Figure 1).

2. Parallel processing strategies

Display technologies *in vitro* are based on the same « idea »: the creation of large diverse libraries of proteins followed by their interrogation using display technologies *in vitro*. An antibody fragment (single-chain Fv (scFv), camelids single domain antibodies (VHH) or Fab fragments) (Figure 2) or an enzyme can be presented on phage or yeast surfaces as well as on ribosomes, while the encoding nucleotide sequence is incorporated within or is physically attached.

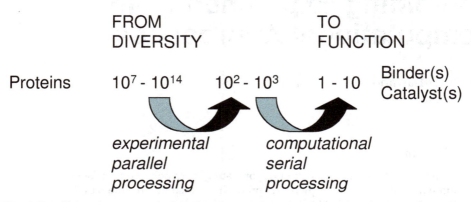

Fig. 1. Parallel and experimental processing combined with serial and computational processing prior to thermodynamic and kinetic characterization allow protein engineering towards new functions.

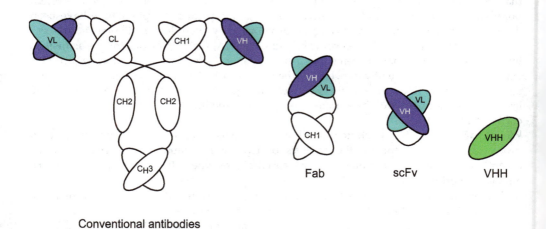

Fig. 2. Representation of mammalian antibodies and synthetic fragments: Fab, scFv and VHH.

This link of phenotype to genotype enables selection and enrichment of molecules with high specific affinities or exquisite catalytic properties together with the co-selected gene (Figure 3). Consequently, the need for serial screening is reduced to a minimum.

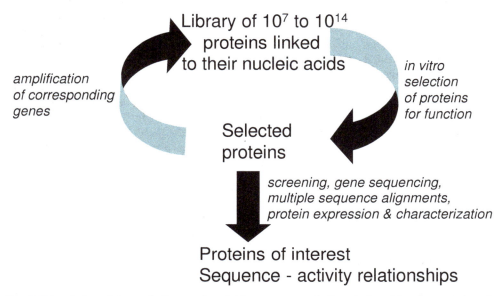

Fig. 3. Directed protein evolution cycles yield sequence-activity relationships for proteins. A cycle consists of the selection of proteins according to their function and of the amplification of their corresponding nucleic acids which are linked to the proteins. Iteration of the cycles diminishes the background of the selection and yields a selected population enriched in proteins with functions of interest. Characterization of these selected proteins and their genes establishes sequence-activity relationships.

2.1 Phage display

In 1985, M13 phage displaying a specific peptide antigen on its surface was isolated from a population of wild type phage, based on the affinity of a specific antibody for the peptide (Smith, 1985). Antibody variable domain were successfully displayed by McCafferty et al in 1990, enabling the selection of antibodies themselves (McCafferty et al., 1990) (Figure 4).

Fig. 4. Bacteriophage particle highlighting the link between a protein fused to a phage coat protein and its corresponding gene located on the phagemid. In the case of *Inovirus*, the filamentous phage particle is a cylinder with a diameter of three to five nanometers, which is about one micrometer long.

Phage display technology (Figure 4) enables the selection from repertoires of antibody fragments (scFv, Fab, VHH) displayed on the surface of filamentous bacteriophage (Smith, 1985). VHH domains are displayed by fusion to the viral coat protein, allowing phage with antigen binding activities (and encoding the antibody fragments) to be selected by panning on antigen. The selected phage can be grown after each round of panning and selected again, and rare phage ($< 1/10^6$) isolated over several rounds of panning.

The antibody fragments genes population is first isolated from lymphocytes then converted to phage-display format using PCR. The PCR products are digested and ligated into phage vector. Subsequent transformation usually yield libraries of 10^6 to 10^{11} clones, each clone corresponding to a specific antibody fragments (VHH, scFv, Fab). This library is panned against the antigen then expression of selected clones is performed. Their biochemical characteristics are analyzed (purity, affinity, specificity) as well as their biological characteristics.

The major advantages of phage display compared with other display technologies are its robustness, simplicity, and the stability of the phage particles, which enables selection on cell surfaces (Ahmadvand et al., 2009), tissue sections (Tordsson et al., 1997) and even *in vivo* (Pasqualini & Ruoslahti, 1996). However, because the coupling of genotype and phenotype (i.e. protein synthesis and assembly of phage particles) takes place in bacteria, the encoding DNA needs to be imported artificially. Library size is therefore restricted by transformation efficiency. Despite great improvements in this area, the largest reported libraries still comprise no more than 10^{10} to 10^{11} different members. Moreover, the amplification of selected variants *in vivo* can lead to considerable biases. Antibody fragments that are toxic for the host, poorly expressed or folded, inefficiently incorporated into the phage particle or susceptible to proteolysis or aggregation slow down the bacterial growth and display less efficiently. This reduces the library diversity and enables a low potency but fast growing clone to dominate a whole population after just a few rounds of selection.

2.2 Ribosome display

Ribosome display was first developed by Dower *et al* (Mattheakis et al., 1994) where mRNA, ribosome and correctly folded functional peptide in a linked assembly could be used for screening and selection (Figure 5).

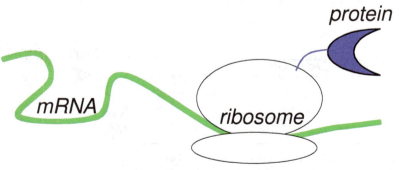

Fig. 5. Ribosome providing the link between the nascent protein and the corresponding mRNA. Such complexes are stabilized in the absence of stop codons and at low temperatures.

In ribosome display, a DNA library coding for particular proteins, for instance scFv or VHH fragments of antibodies, is transcribed *in vitro*. The mRNA is purified and used for *in vitro* translation. Because the mRNA lacks a stop codon, the ribosome stalls at the end of the mRNA, giving rise to a ternary complex of mRNA, ribosome and functional protein. A library of these ternary complexes is tested against the potential ligand (in the case of the antibodies, against the antigen). The binding of the ternary complexes (ribosome, mRNA and protein) to the ligand allows the recovery of the encoding mRNA that is linked to it and that can be transcribed into cDNA by reverse transcriptase-PCR (RT-PCR). Cycles of selection and recovery can be iterated both to enrich rare ligand-binding molecules, and to select molecules with the best affinity.

Ribosome display has been used for the selection of proteins, such as scFv antibody fragments and alternative binding scaffolds with specificity and affinity to peptides (Hanes & Pluckthun, 1997), proteins (Hanes et al., 2000; Knappik et al., 2000; Binz et al., 2004; Lee et al., 2004; Mouratou et al., 2007) and nucleic acids (Schaffitzel et al., 2001). Using transition-state analogs or enzyme inhibitors that bind reversibly to their enzyme (suicide substrates), ribosome display can also be used for the selection for enzymatic activity.

As it is entirely performed *in vitro*, there are two main advantages over other selection strategies. First, the diversity is not limited by the transformation efficiency of bacterial cells, but only by the number of ribosomes and different mRNA molecules present in the test tube. According to the fact that the functional diversity is given by the number of ribosomal complexes that display a functional protein, this number is limited by the number of functional ribosomes or different mRNA molecules, whichever is smaller. An estimate representing a lower limit, of the number of active complexes with folded protein was determined as 2.6×10^{11} per milliter of reaction (Zahnd et al., 2007) and probably is about 10^{13}. Second, random mutations can be introduced easily after each selection rounds, as no library must be transformed after any diversification steps. This allows facile directed evolution of binding proteins over several generations.

However, ribosome display suffers some drawbacks because RNA is extremely labile to ubiquitous Rnases, because the ternary RNA-ribosome-protein complex is very sensitive to heat denaturation and to salt concentration and because large proteins such as DNA polymerases cannot necessarily be produced by *in vitro* translation.

2.3 Yeast surface display

Yeast surface display (YSD) was first demonstrated as a method to immobilize enzymes and pathogen-derived proteins for vaccine development. The β-galactosidase gene from *Cyamopsis tetragonoloba* was fused to the C terminal half of α-agglutinin, a cell wall anchored mating protein in *S. cerevisiae* (Schreuder et al., 1993).

Increased stability was seen for the enzyme when linked to the cell wall, compared with direct secretion of the full β-galactosidase enzyme into the media. Early work also used the flocculin Flo1p as an anchor to attach β-galactosidase to the cell wall, with similar results (Schreuder et al., 1996). Both α-agglutinin and flocculin, along with cell wall proteins such as Cwp1p, Cwp2p, Tip1p, and others, belong to the glycosylphosphatidylinositol (GPI) family of cell wall proteins that can be used directly for display (Kondo & Ueda, 2004). These proteins are directed to the plasma membrane via GPI anchors and subsequently are linked directly to the cell wall through a β-1,6-glucan bridge for incorporation into the mannoprotein layer (Kondo & Ueda, 2004). These large intact proteins as well as their C-terminal fragments have been demonstrated to mediate display of a range of heterologous proteins upon protein fusion.

The α-agglutinin system developped by Wittrup et al (Boder & Wittrup, 1997; Boder et al., 2000; Boder & Wittrup, 2000) uses Aga2p as the display fusion partner. A disulfide linkage between Aga1p, a GPI/β-1,6-glucan-anchored protein, and Aga2p anchors the protein to the cell wall. Thus, coexpression of Aga1p with an Aga2p fusion leads to cell wall-anchored protein on the surface of yeast via disulfide bonding. The majority of applications of YSD utilize now the Aga2p anchor system.

In the yeast surface display system (Figure 6), the antibody fragment (scFv for example) is fused to the adhesion subunit of the yeast agglutinin protein Aga2p, which attaches to the yeast cell wall through disulfide bonds to Aga1p. Each yeast cell typically displays 1.10^4 to 1.10^5 copies of the scFv, and variations in surface expression can be measured through immuno-fluorescence labeling of either the hemagglutinin or c-Myc epitope tag flanking the scFv.

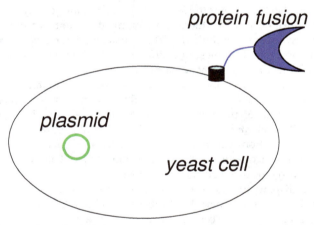

Fig. 6. Yeast cell providing the link between a protein of interest fused to a membrane protein and its corresponding gene located on a plasmid.

In the yeast surface display system (Figure 6), the antibody fragment (scFv for example) is fused to the adhesion subunit of the yeast agglutinin protein Aga2p, which attaches to the yeast cell wall through disulfide bonds to Aga1p. Each yeast cell typically displays 1.10^4 to 1.10^5 copies of the scFv, and variations in surface expression can be measured through immuno-fluorescence labeling of either the hemagglutinin or c-Myc epitope tag flanking the scFv. Likewise, binding to a soluble antigen of interest can be determined by labeling of yeast with biotinylated antigen and a secondary reagent such as streptavidin conjugated to a fluorophore. The display of scFv antibody fragments on the surface of yeast is a powerful and robust technique for the selection of affinity reagents (van den Beucken et al., 2003). Using yeast display for probing immune libraries offers one major advantages over alternative systems. The main advantage is that the scFv displaying yeast can be isolated by FACS and characterized by flow cytometry. The use of FACS in the selection procedure allows the visualization of antigen binding in real-time and the enrichment of each step in the selection can be easily quantified using statistical analyses. Modern flow cytometers can

easily screen millions of binding events in only a few minutes, and the precision of sorting antigen-binding yeast while eliminating nonspecific interactions facilitates large enrichments in a relatively short period of time. In addition, following selection of scFv clones, YSD allows the determination of steady-state kinetic parameters by flow cytometry (K_D value determination (VanAntwerp & Wittrup, 2000)).
However, current yeast display technology is limited by the size of libraries that can be generated and, typically, only libraries of between 10^6 and 10^7 mutants are routinely possible using conventional *in vitro* cloning and transformation.

3. Binders analyzed by parallel processing

This chapter will focus on antibodies, the major class of known binding proteins.

3.1 Introduction on the natural diversity of immunoglobulins

One characteristic of the immune response in vertebrate is the possibility to raise immunoglobulin (Ig) against any type of antigen (Ag), known or unknown. An Ig contains two regions: the Variable domain involved in the binding with the Ag and the Constant domain with effector functions. Each Ig is unique and the variable domain, which is present in each heavy and light chain of every antibody, differ from one antibody to an other. Differences between the variable domains are located on three loops known as complementarity determining regions CDR1, CDR2 and CDR3. CDRs are supported within the variable domains by conserved framework regions. The variability of Ig is based on two phenomena: somatic recombination and somatic hypermutation (SHM).
Somatic recombination of Ig, also known as V(D)J recombination, involves the generation of a unique Ig variable region. The variable region of each immunoglobulin heavy or light chain is encoded in several gene segments. These segments are called variable (V), diversity (D) and joining (J) segments. V, D and J segments are found in Ig heavy chains, but only V and J segments are found in Ig light chains. The IgH locus contains up to 65 VH genes, 27 D genes and 6 J genes while the IgL locus contains 40 V genes and 4-5 J genes, knowing that there are two light chains kappa and lambda. In the bone marrow, each developing B cell will assemble an immunoglobulin variable region by randomly selecting and combining one V, one D and one J gene segment (or one V and one J segment in the light chain). For heavy-chains there are about 10530 potential recombinations (65x27x6) and for light chains 360 potential recombinations (200+160). Moreover some mutations (referred as N-diversity somatic mutations) occur during recombination increasing the diversity by a factor 10^3. These two phenomena, recombination and somatic mutations, lead to about 10^6-10^7 possibilities for heavy chains and $3.5 \; 10^5$ possibilities for light chains generating the formation of about 2.10^{12} different antibodies and thus different antigen specificities (Figure 7) (Jones & Gellert, 2004).
Following activation with antigen, B cells begin to proliferate rapidly. In these rapidly dividing cells, the genes encoding the variable domains of the heavy and light chains undergo a high rate of point mutation, by a process called somatic hypermutation (SHM). The SHM mutation rate is about 10^{-3} per base pair and per cell division, that is approximately one million times above the replicative mutation rate. As a consequence, any daughter B cells will acquire slight amino acid differences in the variable domains of their antibody chains.

This serves to increase the diversity of the antibody pool and impacts the antibody's antigen-binding affinity. Some point mutations will result in the production of antibodies that have a weaker interaction (low affinity) with their antigen than the original antibody, and some mutations will generate antibodies with a stronger interaction (high affinity). It has been estimated that the affinity of an immunoglobulin for an antigen is raised by a factor 10 to 100 (Kepler & Bartl, 1998). B cells that express high affinity antibodies on their surface will receive a strong survival signal during interactions with other cells, whereas those with low affinity antibodies will not, and will die by apoptosis. The process of generating antibodies with increased binding affinities is called affinity maturation (Neuberger, 2008).

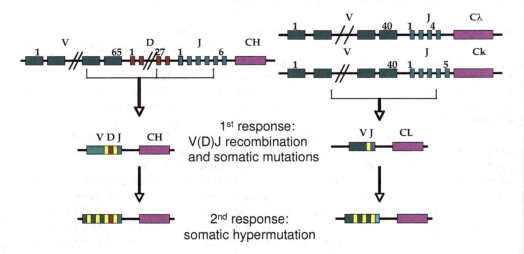

Fig. 7. Recombination and hypermutation of immunoglobulins. A yellow rectangle represents a point mutation. Recombination and somatic hypermutation are shown for heavy chains (left) and for light chains (right).

This quite complex process for generation of highly specific antibodies is a challenge for the obtention of recombinant antibodies. Many factors influence the quality of the recombinant antibodies: starting or not from an immunized animals or humans, the size and the quality of the libraries, the possibility to mutate the antibodies.

3.2 Antibody libraries
3.2.1 Recombinant antibody libraries
Recombinant antibody libraries have been constructed by cloning antibody heavy- or light-chain variable genes directly from lymphocytes of animals or human and then expressing as a single-chain fragment variable (scFv) single-domain antibodies (VHH) or as an antigen-binding fragment (Fab) using various display technologies. The recombinant antibody technology, an alternative to traditional immunization of animals, facilitates to isolate target specific high affinity monoclonal antibodies without immunization by virtue of combination with high throughput screening techniques.

A strategy for creation of a combinatorial antibody library is very important to isolate high specificity and affinity antibodies against target antigens. To date, a variety of different antibody libraries have been generated, which range from immune to naive and even synthetic antibody libraries (Table 1). Immune libraries derived from IgG genes of immunized donors (Sanna et al., 1995) are useful if immunized patients are available but have the disadvantage that antibodies can only be made against the antigens used for immunization. In contrast, antibodies against virtually any antigen, including self-, non-immunogenic, or toxic antigens, can be isolated from naive or synthetic libraries. Naive libraries from non-immunized donors have been generated by PCR-cloning Ig repertoires from various B-cell sources (Marks et al., 1991; Vaughan et al., 1996; Sheets et al., 1998; de Haard et al., 1999)) derived from human or camel germ line genes and randomized only in the CDR3 regions (Hoogenboom & Winter, 1992; Nissim et al., 1994; de Kruif et al., 1995). Synthetic libraries have been generated from a repertoire of 49 human germline VH genes segments rearranged *in vitro* to create a synthetic CDR3 region (Hoogenboom & Winter, 1992) or derived from a single V-gene with complete randomization of all CDRs (Jirholt et al., 1998; Soderlind et al., 2000) (Table 1).

	Synthetic	Naive	Immune
V-gene source	Unrearranged V-gene segments	Rearranged-V genes from Ig pool	Rearranged V-genes from specific IgG pool
Contents	controlled	uncontrolled	uncontrolled
Repertoire construction	Once (single pot)		New repertoire for every antigen
Affinity of antibodies	Depending on library size : μM from standard size repertoire (10^7) nM from very large repertoire (10^{10})		Biased for high affinity (nM if antigen is immunogenic)
specificity	Any	Originally biased against self	Immunodominant epitopes, biased against self

Table 1. Comparison between Synthetic, Naive and Immune libraries (according to (Hoogenboom, 1997))

3.2.2 Immune libraries

Efficient isolation of specific high affinity binders from relatively small sized libraries was shown using immune antibody libraries constructed from B lymphocytes of immunized mice, camels or patients with a high antibody titer for particular antigens, in our laboratory and by others: a targeted immune library contained typically about 10^6 clones (Burton, 1991; Barbas et al., 1992; Barbas et al., 1992) (Table 1). However, the construction of an immune library is not always possible due to the difficulty in obtaining antigen-related B lymphocytes.

The quality of the immune response will likely dictate the outcome of the library selections. It is generally accepted that early in the immune response the repertoire of immunoglobulins is diverse and of low affinity to the antigen. The process of SHM through successive rounds of selection ensures that the surviving B cells develop progressively higher affinities, but probably at the expense of diversity. The balance between diversity and affinity is something that may be exploited by researchers depending on the goal of their study.

3.2.3 Non-immune libraries

The most important parameter in the non-immune antibody library is the library diversity, i.e., library size, in an aspect that, in general, the larger the library, the higher the likelihood is to isolate high affinity binders to a particular antigen. Typically, a 10^9 to 10^{11} library diversity has been reported to generate specific high affinity binders with dissociation constants in the 1–1000 nM range (Table 1). For example, scFvs against crotoxin, a highly toxic-β-neurotoxin isolated from the venom of the rattlesnake, *Crotalus durrissus terrificus*, have been selected from two non-immune scFv libraries which differ by their size; respectively 10^6 (Nissim et al., 1994) and 10^{10} diversity (Vaughan et al., 1996). The affinity of anti-crotoxin scFvs is in the micromolar range in the first case and in the nanomolar range in the second case. Moreover, these latter scFvs possessed an *in vivo* neutralizing activity against a venom toxin.

However, creating a large antibody library is time consuming and does not always guarantee to isolate high affinity binders to any given antigen. Therefore, many attempts have been undertaken to make the library size as big as possible, and site-specific recombination systems have been created to overcome the library size limitations given by the conventional cloning strategies. Besides library generation, the panning process itself limits also the library size that can be handled conveniently.

Therefore, it is important to generate libraries with a high quality of displayed antibodies, thus emphasizing the functional library size and not only the apparent library size. For instance, one limitation of phage display is that it requires prokaryotic expression of antibody fragments. It is well known that there is an unpredictable expression bias against some eukaryotic proteins expressed from *Escherichia coli* because the organism lacks foldases and chaperones present in the endoplasmic reticulum of eukaryotic cells that are necessary for efficient folding of secreted proteins such as antibody fragments. Even minor sequence changes such as single point mutations in the complementarity determining regions (CDRs) of Fab fragments can completely eliminate antibody expression in *E. coli* (Ulrich et al., 1995), and a random sampling of a scFv library showed that half of the library had no detectable level of scFv in the culture supernatant (Vaughan et al., 1996). Because the protein folding and secretory pathways of yeast more closely approximate those of mammalian cells, it has been shown that yeast display could provide access to more antibodies than phage display (Bowley et al., 2007). In this study, the two approaches were directly compared using the same HIV-1 immune scFv cDNA library expressed in phage and yeast display vectors and using the same selecting antigen (HIV-1 gp120). After 3 to 4 rounds of selection, sequence analysis of individual clones revealed many common antibodies isolated by both techniques, but also revealed many novel antibodies derived from the yeast display selection that had not previously been described by phage display. It appears that the level of expression of correctly folded scFv on the phage surface is one of the most important criteria for selection.

VHH libraries may be an advantageous alternative because VHH are highly soluble, stable, easily expressed in *E. coli* and because they do not tend to aggregate (Muyldermans, 2001; Harmsen & de Haard, 2007). Moreover due to their small size (15 kDa compared to 25-30 kDa for a scFv and 50 kDa for a Fab), VHH could diffuse easily in tissues, bind to poorly accessible epitopes for conventional antibody fragments (Desmyter et al., 1996; Stijlemans et al., 2004) and bind non-conventional epitopes (Behar et al., 2009). Chen *et al* (Chen et al., 2008) have prepared a phage displayed VH-only domain library by grafting naturally occurring CDR2 and CDR3 of heavy chains on a VHH-like scaffold. From this library (size $2.5 \ 10^{10}$) they have selected high quality binders against viral and cancer-related antigens. From a non-immune VHH library of 10^8 diversity, VHH have been selected against various viral protein by phage display. These VHH had an affinity in the nanomolar range but more interestingly the k_{off} is very low (about 10^4 to 10^5 s^{-1}) allowing them to be suitable for crystallographic studies (Lafaye –personnal communication).

3.3 Affinity optimization
With tools such as phage, yeast and ribosome display available to isolate rapidly specific high-potency antibodies from large variant protein populations, a major key to efficient and successful *in vitro* antibody optimization is the introduction of the appropriate sequence diversity into the starting antibody. Generally, two approaches can be taken: either amino acid residues in the antibody sequence are substituted in a targeted way or mutations are generated randomly.

3.3.1 Affinity increase by targeted mutations
Antibodies are ideal candidates for targeted sequence diversification because they share a high degree of sequence similarity and their conserved immunoglobulin protein fold is well studied. Many *in vitro* affinity maturation efforts using combinatorial libraries in conjunction with display technologies have targeted the CDRs harbouring the antigen-binding site. Normally, amino acid residues are fully randomized with degenerate oligonucleotides. If applied to all positions in a given CDR, however, this approach would create far more variants than can be displayed on phage, on yeast or even ribosomes – saturation mutagenesis of a CDR of 12 residues, for example, would result in 20^{12} different variants. In addition, the indiscriminate mutation of these residues creates many variants that no longer bind the antigen, reducing the functional library size. Scientists have therefore restricted the number of mutations by targeting only blocks of around six consecutive residues per library (Thom et al., 2006) or by mutating four variants in all the CDRs (Laffly et al., 2008) or by mutating only the CDRs 1 and 2 (Hoet et al., 2005). Mutagenesis has also been focussed on natural hotspots of SHM (Ho et al., 2005). In other works, the residues to be targeted were chosen based on mutational or structural analyses as well as on molecular models (Yelton et al., 1995; Osbourn et al., 1996; Chen & Stollar, 1999). Further affinity improvements have been achieved by recombining mutations within the same or different CDRs of improved variants (Jackson et al., 1995; Yelton et al., 1995; Chen & Stollar, 1999; Rajpal et al., 2005). Despite some substantial gains, such an approach is unpredictable. As an alternative, CDRs were sequentially mutated by iterative constructions and pannings of libraries, starting with CDR3, in a strategy named « CDR walking » (Yang et al., 1995; Schier et al., 1996). Although this results in greater improvements, it is time consuming and permits only one set of amino acid changes to recombine with new mutations.

3.3.2 Affinity increase by random mutations
In addition to the targeted strategies, several random mutagenesis methods can be used to improve antibody potency. One is the shuffling of gene segments, where VH and VL populations, for example, can be randomly recombined with each other (Figini et al., 1994;

Schier et al., 1996) or be performed with CDRs (Jirholt et al., 1998; Knappik et al., 2000). An alternative approach is the possibility that independent repertoires of heavy chain (HC) and light chain (LC) can be constructed in haploid yeast strains of opposite mating type. These separate repertoires can then be combined by highly efficient yeast mating. Using this approach, Blaise *et al* (Blaise et al., 2004) have rapidly generated a human Fab yeast display library of over 10^9 clones, allowing the selection of high affinity Fab by YSD using a repeating process of mating- driven chain shuffling and flow cytometric sorting.

Another approach is the indiscriminate mutation of nucleotides using the low-fidelity Taq DNA polymerase (Hanes et al., 2000), error-prone PCR (Hawkins et al., 1992; Daugherty et al., 2000; Jermutus et al., 2001; van den Beucken et al., 2003), the error-prone Qbeta RNA replicase (Irving et al., 2001) or *E. coli* mutator strains (Irving et al., 1996; Low et al., 1996; Coia et al., 2001) before and in-between rounds of selection. Shuffling and random point mutagenesis are particularly useful when used in conjunction with targeted approaches because they enable the simultaneous evolution of non-targeted regions (Thom et al., 2006); in addition, they are powerful when performed together because individual point mutations can recombine and cooperate, again leading to synergistic potency improvements. This has created some of the highest affinity antibodies produced so far, with dissociation constants in the low picomolar range (Zahnd et al., 2004) and in a study using yeast display, even in the femtomolar range (Boder et al., 2000). When performed separately, random mutagenesis can help identify mutation hotspots, defined as amino acid residues mutated frequently in a population. To this end, a variant library generated by error-prone PCR, for example, might be subjected to affinity selections followed by the sequencing of improved scFvs. In a manner similar to somatic hypermutation, this method leads to the accumulation of mutations responsible for potency gains mainly in CDRs, despite having been introduced randomly throughout the whole scFv coding sequence (Thom et al., 2006).

3.3.3 Affinity increase by selection optimization

Mutant libraries are often screened under conditions where the binding interaction has reached equilibrium with a limiting concentration of soluble antigen to select mutants having higher affinity. When labelled with biotin, for example, the antigen and the bound scFv-phage, scFv-yeast or scFv-ribosome-mRNA complexes can be pulled down with streptavidin-coated magnetic beads. The antigen concentration chosen should be below the K_D of the antibody at the first round of selection and then reduced incrementally during subsequent cycles to enrich for variants with lower K_D (Hawkins et al., 1992; Schier et al., 1996). Selections have been performed in the presence of an excess of competitor antigen or antibody, resulting specifically in variants with lower off-rates (Hawkins et al., 1992; Jermutus et al., 2001; Zahnd et al., 2004; Laffly et al., 2008).

Protein affinity maturation has been one of the most successful applications of YSD. Initial studies led by Wittrup *et al.* used an anti-fluorescein scFv to show the effectiveness of YSD in protein affinity maturation (Boder et al., 2000; Feldhaus & Siegel, 2004). Since each yeast cell is capable of displaying 10^4 to 10^5 copies of a single scFv (Boder & Wittrup, 1997), fluorescence from each cell can be readily detected and accurately quantified by flow cytometry. This feature of YSD allows not only precise and highly reproducible affinity measurement, but also rapid enrichment of high-affinity populations within mutant libraries (Boder et al., 2000). Moreover, on-rate selections have been realized only with yeast display, which profits from using flow cytometric cell sorting to finely discriminate variants with specified binding kinetics (Razai et al., 2005).

The selected antibodies can be tested for increased affinity but should preferentially be screened for improved potency in a relevant cell-based assay because the sequence diversification and selection process might also have enriched variants with increased folding efficiency and thermodynamic stability, both contributing to potency and, ultimately, efficacy.

3.4 Conclusions on the parallel processing of binders

Phage, yeast and ribosome display were proven to be powerful methods for screening libraries of antibodies. By means of selection from large antibody repertoires, a wide variety of antibodies have been generated in the form of scFv, VHH or Fab fragments. After a few rounds of panning or selection on soluble antigens and subsequent amplification in *E. coli*, large numbers of selected clones have to be analyzed with respect to antigen specificity, and binding affinity. Analysis of these selected binders is usually performed by ELISA. Hopefully, the introduction of automated screening methods to the display process provides the opportunity to evaluate hundreds of antibodies in downstream assays. Secondary assays should minimally provide a relative affinity ranking and, if possible, reliable estimates of kinetic or equilibrium affinity constants for each of the hits identified in the primary screen.

Surface plasmon resonance (SPR) methods has been used to measure the thermodynamic and kinetic parameters of antigen-antibody interactions. An SPR secondary screening assay must be capable of rapidly analyzing all the unique antibodies discovered in the primary screen. The first generations of widely used commercial systems from Biacore process only one sample at a time and this limits the throughput for antibody fragments screening to approximately 100 samples per day. Recently however, several biosensors were introduced to increase the number of samples processed with different approaches for sample delivery (Wassaf et al., 2006) (Safsten et al., 2006; Nahshol et al., 2008).

To reduce the number of antibodies tested and so far the amount of antigen used, it is crucial to analyze the diversity of the antibody fragments after the first screening performed by ELISA. Usually after few rounds of selection, a limited number of clones, found in several copies, are obtained. In that case, it is un-necessary to analyze such redundant clones. It is the reason why we have decided in our laboratory to sequence the clones after the first screening, then to analyze only the unique clones by SPR in a secondary screening.

Despite the growing knowledge around antibody structures and protein–protein interactions, and the rapid development of *in silico* evolution, molecular modelling and protein–protein docking tools, it is still nearly impossible to predict the multitude of mutations resulting in improved antibody potency. Moreover, specific structural information – on the antibody to be optimized (paratope), its antigen (epitope) and their interaction – can lack the high resolution required to determine accurately important details such as side-chain conformations, hydrogen-bonding patterns and the position of water molecules. Therefore, the most effective way to improve antibody potencies remains the use of display technologies to interrogate large variant populations, using either targeted or random mutagenesis strategies.

4. Catalysts analyzed by parallel processing

4.1 Enzyme libraries

To isolate rare catalysts of interest for specific chemical reactions, the parallel processing of millions of mutant enzymes turned out to be a successful strategy (Figures 3&8). Various types of protein libraries can be constructed. Almost random protein sequences have been designed and submitted to selection for the isolation of nucleic acid ligases (Seelig & Szostak, 2007). Given that most enzymes have more than 50 amino acids, and that each amino acid can be one out of twenty in the standard genetic code, 20^{50} distinct sequences can be considered. The parallel or serial processing of so many proteins cannot be conceived experimentally. A useful strategy then relies on the directed evolution of known enzymes, which catalyze chemical reactions that are similar to the reactions of interest (Figure 8).

Enzyme libraries have been constructed by random mutagenesis of the corresponding genes. This can be achieved by introduction of manganese ions within PCR mixtures during amplification of the gene encoding the enzyme. Manganese ions alter the fidelity of the DNA-dependent DNA polymerase used for amplification and provided their concentration is precisely adjusted, the average number of base substitutions per gene can be accurately evaluated (Cadwell & Joyce, 1994). Concentrations of deoxynucleotides triphosphates can be further adapted so as to define the relative rates of different base substitutions (Fromant et al., 1995).

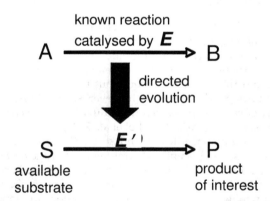

Fig. 8. Strategy for the isolation of a catalyst of interest E' for the reaction from S to P by directed evolution of an enzyme E catalyzing a similar chemical reaction converting A into B.

Other enzyme libraries have been constructed by directed mutagenesis at specific sites within proteins, for example in our laboratory. Known x-ray crystal structures of enzymes in complex with their substrates can be used as a basis to identify the specific amino acids known to bind the substrates at the active site (Figure 9).

Oligonucleotides can be further synthesized with random mutations introduced specifically at the very few codons coding amino acids known to interact with the substrates. PCR assembly of such oligonucleotides can then be used to reconstitute full-length open reading frames coding for mutant proteins. Experience from our laboratory indicates that protein libraries designed by introduction of quasi-random mutations over an entire protein domain yield a higher number of catalysts of interest than protein libraries carefully designed by introduction of mutations at specific sites within the active site. This strategy requires nevertheless an efficient parallel processing strategy for analysis of millions of protein mutants.

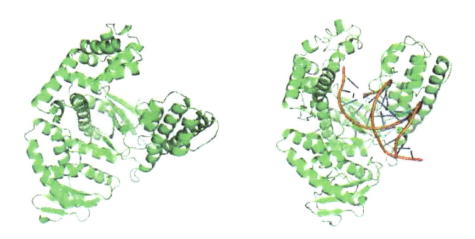

Fig. 9. Comparison of *Thermus aquaticus* DNA polymerase I's Stoffel fragment structures with (2ktq) and without a DNA duplex (1jxe) at the active site.

4.2 Selections from enzyme libraries

Design of selections for the isolation of catalysts from large protein repertoires has been far from obvious. The various parallel processing strategies to identify active enzymes rely generally on selections for binding. Selections for binding to suicide inhibitors were first tested (Soumillion et al., 1994). Selection of protein mutants for binding to transition state analogues yield in principle catalysts. This approach remains delicate, possibly because of the rough similarity between transition states and transition state analogues whose stability is required for the selections, and because of the time required to synthesize transition state analogues by organic synthesis. Successful parallel processing strategies for the isolation of catalysts relied on the selection of multiple products bound to the enzyme complex that catalyzed the chemical reaction. These *in vitro* selections are furthermore selections for the highest catalytic turnovers (Figure 10). Populations of enzymes with the highest catalytic efficiencies are thereby isolated.

Sequencing of the genes encoding hundred variants of the selected population then allows multiple sequence alignments to be carried out for the identification of recurrent mutations which characterize the catalytic activity change or improvement. Further characterization of isolated catalysts consists of the measurement of the kinetic parameters for the chemical reactions studied. Improvements of the catalytic efficiencies by several orders of magnitude have been described in the literature for several enzymes. These results have important applications in the field of biocatalysis.

Alternatively, for substrate-cleaving reaction, the concept of catalytic elution was reported (Pedersen et al., 1998): complexes between enzymes displayed on the surface of bacteriophages and their substrates bound to a solid phase are formed. Activation of the enzyme results in release of the phage-enzyme from the solid phase if the enzyme is active, while inactive enzymes remain bound to the solid phase (Soumillion & Fastrez, 2001).

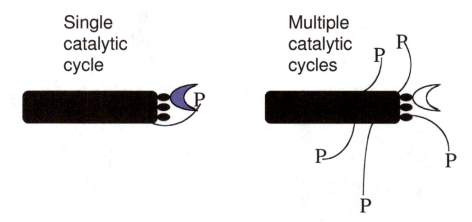

Fig. 10. Comparison of a highly active enzyme (white) efficiently captured by affinity chromatography for the product with a protein catalyzing a single substrate to product conversion (blue) unlikely to be isolated by affinity chromatography for the product.

4.3 Conclusion for enzymes

The parallel processing of molecular information on the catalytic activity of proteins (« Is the protein a catalyst or not ? ») is remarkably achieved by *in vitro* selection from large libraries of millions or billions of mutant proteins. Reduction of the large diversity into a small diversity of hundred(s) of variant proteins with the catalytic activity of interest allows characterization by serial processing to be accomplished. The sequencing of the corresponding genes for hundred(s) of variants allows computation of alignments for multiple sequences. The yield of protein production and the catalytic efficiencies for tens of selected variants allow the most promising variant protein to be identified. These results define sequence-activity relationships for enzymes. If enzyme-substrate complex structures are available, the sequence-structure-activity relationships that can be derived provide the central information for use in further biocatalytic applications.

5. Conclusion

Molecular biology, bioinformatics and protein engineering reached in the last decades a state allowing the isolation of proteins for desired functions of interest. Proteins can be isolated with a binding specificity for a given target, while enzymes can be isolated for given chemical reactions. Binding proteins and antibodies in particular found remarkable applications in the field of therapeutics. Enzymes turn out to be extremely useful in the field of biocatalysis for the production of chemicals at industrial scales within a sustainable environment.

Over the last twenty years, the use of antibodies has increased greatly, both as tools for basic research and diagnostics, and as therapeutic agents. This has largely been driven by ongoing advances in recombinant antibody technology. Today, more than 20 recombinant antibodies are widely used in clinic such as the human anti-TNFα antibody marketed as Humira® and many more antibodies are currently in clinical trials.

Satisfying industrial needs in the field of biocatalysis requires efficient enzymes to be isolated. While natural enzymes rarely fulfill industrial needs, and as long as computational approaches alone do not allow the sequences of protein catalysts to be designed, experimental methods such as the parallel processing strategies relying on *in vitro* selection combined with computational approaches for the characterization of catalysts may well be the most powerful strategies for the isolation of enzymes for given chemical reactions. Most notably, these new biocatalysts act in aqueous solutions without organic solvents at large scale and are ideally suited for green industrial processes.

A highly efficient design of binders and catalysts according to function can make use of a unique strategy: selection from large repertoires of proteins according to a function yield secondary protein repertoires of high interest, which can then be processed in series for their characterization due to their reduced diversity. Characterization involves sequencing of the corresponding genes for alignment of numerous protein sequences so as to define consensus sequences. This is the major advantage of molecular information parallel processing (MIPP) strategies: *defining conserved amino acids within protein scaffolds tightly linked to function.*

In conclusion, the parallel processing of biomolecular information (« Does the protein bind the target ? » or « Is the protein a catalyst for the chemical reaction ? ») is so far best achieved experimentally by using repertoires of millions or billions of proteins. Analysis of hundred(s) of protein variants is then best done computationally: use of multiple sequence alignment algorithms yields then sequence-activity relationships required for protein applications. Further biochemical and biophysical characterization of proteins (« Does the protein tend to form dimers or to aggregate ? », « Can the protein be produced at high level ? » , « What is the protein's pI ? ») is essential for their final use which may require high level soluble expression or cell penetration properties. In this respect, the development of algorithms analyzing protein properties remains a major challenge.

6. References

Ahmadvand, D., Rasaee, M. J., Rahbarizadeh, F., Kontermann, R. E. & Sheikholislami, F. (2009). Cell selection and characterization of a novel human endothelial cell specific nanobody. *Molecular Immunology,* Vol. 46, No. 8-9, pp. 1814-1823.

Barbas, C. F. d., Bjorling, E., Chiodi, F., Dunlop, N., Cababa, D., Jones, T. M., Zebedee, S. L., Persson, M. A., Nara, P. L., Norrby, E. & Burton, D. R. (1992). Recombinant human Fab fragments neutralize human type 1 immunodeficiency virus in vitro. *Proceedings of the National Academy of Sciences USA,* Vol. 89, No. 19, pp. 9339-9343.

Barbas, C. F. d., Crowe, J. E., Jr., Cababa, D., Jones, T. M., Zebedee, S. L., Murphy, B. R., Chanock, R. M. & Burton, D. R. (1992). Human monoclonal Fab fragments derived from a combinatorial library bind to respiratory syncytial virus F glycoprotein and neutralize infectivity. *Proceedings of the National Academy of Sciences USA,* Vol. 89, No. 21, pp. 10164-10148.

Behar, G., Chames, P., Teulon, I., Cornillon, A., Alshoukr, F., Roquet, F., Pugniere, M., Teillaud, J. L., Gruaz-Guyon, A., Pelegrin, A. & Baty, D. (2009). Llama single-domain antibodies directed against nonconventional epitopes of tumor-associated carcinoembryonic antigen absent from nonspecific cross-reacting antigen. *Febs Journal,* Vol. 276, No. 14, pp. 3881-3893.

Binz, H. K., Amstutz, P., Kohl, A., Stumpp, M. T., Briand, C., Forrer, P., Grutter, M. G. & Pluckthun, A. (2004). High-affinity binders selected from designed ankyrin repeat protein libraries. *Nature Biotechnology*, Vol. 22, No. 5, pp. 575-582.

Blaise, L., Wehnert, A., Steukers, M. P., van den Beucken, T., Hoogenboom, H. R. & Hufton, S. E. (2004). Construction and diversification of yeast cell surface displayed libraries by yeast mating: application to the affinity maturation of Fab antibody fragments. *Gene*, Vol. 342, No. 2, pp. 211-218.

Boder, E. T., Midelfort, K. S. & Wittrup, K. D. (2000). Directed evolution of antibody fragments with monovalent femtomolar antigen-binding affinity. *Proceedings of the National Academy of Sciences USA*, Vol. 97, No. 20, pp. 10701-10705.

Boder, E. T. & Wittrup, K. D. (1997). Yeast surface display for screening combinatorial polypeptide libraries. *Nature Biotechnology*, Vol. 15, No. 6, pp. 553-557.

Boder, E. T. & Wittrup, K. D. (2000). Yeast surface display for directed evolution of protein expression, affinity, and stability. *Methods Enzymology*, Vol. 328, No. pp. 430-444.

Bowley, D. R., Labrijn, A. F., Zwick, M. B. & Burton, D. R. (2007). Antigen selection from an HIV-1 immune antibody library displayed on yeast yields many novel antibodies compared to selection from the same library displayed on phage. *Protein Engineering Design Selection*, Vol. 20, No. 2, pp. 81-90.

Burton, D. R. (1991). Human and mouse monoclonal antibodies by repertoire cloning. *Tibtech*, Vol. 9, No. pp. 169-175.

Cadwell, R. C. & Joyce, G. F. (1994). Mutagenic PCR. *PCR Methods & Applications*, Vol. 3, No. 6, pp. 136-140.

Chen, W., Zhu, Z., Feng, Y., Xiao, X. & Dimitrov, D. S. (2008). Construction of a large phage-displayed human antibody domain library with a scaffold based on a newly identified highly soluble, stable heavy chain variable domain. *Journal of Molecular Biology*, Vol. 382, No. 3, pp. 779-789.

Chen, Y. & Stollar, B. D. (1999). DNA binding by the VH domain of anti-Z-DNA antibody and its modulation by association of the VL domain. *Journal of Immunology*, Vol. 162, No. 8, pp. 4663-4670.

Coia, G., Hudson, P. J. & Irving, R. A. (2001). Protein affinity maturation in vivo using E. coli mutator cells. *Journal of Immunological Methods*, Vol. 251, No. 1-2, pp. 187-193.

Daugherty, P. S., Chen, G., Iverson, B. L. & Georgiou, G. (2000). Quantitative analysis of the effect of the mutation frequency on the affinity maturation of single chain Fv antibodies. *Proceedings of the National Academy of Sciences USA*, Vol. 97, No. 5, pp. 2029-2034.

de Haard, H. J., van Neer, N., Reurs, A., Hufton, S. E., Roovers, R. C., Henderikx, P., de Bruine, A. P., Arends, J. W. & Hoogenboom, H. R. (1999). A large non-immunized human Fab fragment phage library that permits rapid isolation and kinetic analysis of high affinity antibodies. *Journal of Biological Chemistry*, Vol. 274, No. 26, pp. 18218-18230.

de Kruif, J., Boel, E. & Logtenberg, T. (1995). Selection and application of human single chain Fv antibody fragments from a semi-synthetic phage antibody display library with designed CDR3 regions. *Journal of Molecular Biology*, Vol. 248, No. 1, pp. 97-105.

Desmyter, A., Transue, T. R., Ghahroudi, M. A., Thi, M. H., Poortmans, F., Hamers, R., Muyldermans, S. & Wyns, L. (1996). Crystal structure of a camel single-domain VH antibody fragment in complex with lysozyme. *Nature Structural Biology*, Vol. 3, No. 9, pp. 803-811.

Feldhaus, M. & Siegel, R. (2004). Flow cytometric screening of yeast surface display libraries. *Methods Molecular Biology*, Vol. 263, No. pp. 311-332.

Figini, M., Marks, J. D., Winter, G. & Griffiths, A. D. (1994). In vitro assembly of repertoires of antibody chains on the surface of phage by renaturation. *Journal of Molecular Biology*, Vol. 239, No. 1, pp. 68-78.

Fromant, M., Blanquet, S. & Plateau, P. (1995). Direct random mutagenesis of gene-sized DNA fragments using polymerase chain reaction. *Analytical Biochemistry*, Vol. 224, No. 1, pp. 347-353.

Hanes, J., Jermutus, L. & Pluckthun, A. (2000). Selecting and evolving functional proteins in vitro by ribosome display. *Methods Enzymology*, Vol. 328, pp. 404-430.

Hanes, J. & Pluckthun, A. (1997). In vitro selection and evolution of functional proteins by using ribosome display. *Proceedings of the National Academy of Sciences USA*, Vol. 94, No. 10, pp. 4937-4942.

Harmsen, M. M. & de Haard, H. J. W. (2007). Properties, production, and applications of camelid single-domain antibody fragments. *Applied Microbiology and Biotechnology*, Vol. 77, No. 1, pp. 13-22.

Hawkins, R. E., Russell, S. J. & Winter, G. (1992). Selection of phage antibodies by binding affinity. Mimicking affinity maturation. *Journal of Molecular Biology*, Vol. 226, No. 3, pp. 889-896.

Ho, M., Kreitman, R. J., Onda, M. & Pastan, I. (2005). In vitro antibody evolution targeting germline hot spots to increase activity of an anti-CD22 immunotoxin. *Journal of Biological Chemistry*, Vol. 280, No. 1, pp. 607-617.

Hoet, R. M., Cohen, E. H., Kent, R. B., Rookey, K., Schoonbroodt, S., Hogan, S., Rem, L., Frans, N., Daukandt, M., Pieters, H., van Hegelsom, R., Neer, N. C., Nastri, H. G., Rondon, I. J., Leeds, J. A., Hufton, S. E., Huang, L., Kashin, I., Devlin, M., Kuang, G., Steukers, M., Viswanathan, M., Nixon, A. E., Sexton, D. J., Hoogenboom, H. R. & Ladner, R. C. (2005). Generation of high-affinity human antibodies by combining donor-derived and synthetic complementarity-determining-region diversity. *Nature Biotechnology*, Vol. 23, No. 3, pp. 344-348.

Hoogenboom, H. R. (1997). Designing and optimizing library selection strategies for generating high-affinity antibodies. *Trends Biotechnology*, Vol. 15, No. 2, pp. 62-70.

Hoogenboom, H. R. & Winter, G. (1992). By-passing immunisation. Human antibodies from synthetic repertoires of germline VH gene segments rearranged in vitro. *Journal of Molecular Biology*, Vol. 227, No. 2, pp. 381-388.

Irving, R. A., Coia, G., Roberts, A., Nuttall, S. D. & Hudson, P. J. (2001). Ribosome display and affinity maturation: From antibodies to single V-domains and steps towards cancer therapeutics. *Journal of Immunological Methods*, Vol. 248, No. 1-2, pp. 31-45.

Irving, R. A., Kortt, A. A. & Hudson, P. J. (1996). Affinity maturation of recombinant antibodies using E. coli mutator cells. *Immunotechnology*, Vol. 2, No. 2, pp. 127-143.

Jackson, J. R., Sathe, G., Rosenberg, M. & Sweet, R. (1995). In vitro antibody maturation. Improvement of a high affinity, neutralizing antibody against IL-1 beta. *Journal of Immunology*, Vol. 154, No. 7, pp. 3310-3319.

Jermutus, L., Honegger, A., Schwesinger, F., Hanes, J. & Pluckthun, A. (2001). Tailoring in vitro evolution for protein affinity or stability. *Proceedings of the National Academy of Sciences USA*, Vol. 98, No. 1, pp. 75-80.

Jirholt, P., Ohlin, M., Borrebaeck, C. A. K. & Soderlind, E. (1998). Exploiting sequence space: shuffling in vivo formed complementarity determining regions into a master framework. *Gene*, Vol. 215, No. 2, pp. 471-476.

Jones, J. M. & Gellert, M. (2004). the taming of a transposon: V(D)J recombination and the immune system. *Immunological Review*, Vol. 200, No. 1, pp. 233-248.

Kepler, T. B. & Bartl, S. (1998). Plasticity under somatic mutations in antigen receptors. *Current Topics in Microbiology & Immunology*, Vol. 229, pp. 149-162.

Knappik, A., Ge, L., Honegger, A., Pack, P., Fischer, M., Wellnhofer, G., Hoess, A., Wolle, J., Pluckthun, A. & Virnekas, B. (2000). Fully synthetic human combinatorial antibody libraries (HuCAL) based on modular consensus frameworks and CDRs randomized with trinucleotides. *Journal of Molecular Biology*, Vol. 296, No. 1, pp. 57-86.

Kondo, A. & Ueda, M. (2004). Yeast cell-surface display--applications of molecular display. *Applied Microbiology Biotechnology*, Vol. 64, No. 1, pp. 28-40.

Laffly, E., Pelat, T., Cedrone, F., Blesa, S., Bedouelle, H. & Thullier, P. (2008). Improvement of an antibody neutralizing the anthrax toxin by simultaneous mutagenesis of its six hypervariable loops. *Journal of Molecular Biology*, Vol. 378, No. 5, pp. 1094-1103.

Lee, M. S., Kwon, M. H., Kim, K. H., Shin, H. J., Park, S. & Kim, H. I. (2004). Selection of scFvs specific for HBV DNA polymerase using ribosome display. *Journal of Immunology Methods*, Vol. 284, No. 1-2, pp. 147-157.

Low, N., Holliger, P. & Winter, G. (1996). Mimicking somatic hypermutation: affinity maturation of antibodies displayed on bacteriophage using a bacterial mutator strain. *Journal of Molecular Biology*, Vol. 260, No. 3, pp. 359-368.

Marks, J. D., Hoogenboom, H. R., Bonnert, T. P., McCafferty, J., Griffiths, A. D. & Winter, G. (1991). By-passing immunization. Human antibodies from V-gene libraries displayed on phage. *Journal of Molecular Biology*, Vol. 222, No. 3, pp. 581-597.

Mattheakis, L. C., Bhatt, R. R. & Dower, W. J. (1994). An in vitro polysome display system for identifying ligands from very large peptide libraries. *Proceedings of the National Academy of Sciences USA*, Vol. 91, No. 19, pp. 9022-9026.

McCafferty, J., Griffiths, A. D., Winter, G. & Chiswell, D. J. (1990). Phage antibodies: filamentous phage displaying antibody variable domains. *Nature*, Vol. 348, No. 6301, pp. 552-554.

Mouratou, B., Schaeffer, F., Guilvout, I., Tello-Manigne, D., Pugsley, A. P., Alzari, P. M. & Pecorari, F. (2007). Remodeling a DNA-binding protein as a specific in vivo inhibitor of bacterial secretin PulD. *Proceedings of the National Academy of Sciences USA*, Vol. 104, No. 46, pp. 17983-17988.

Muyldermans, S. (2001). Single domain camel antibodies: current status. *Reviews in Molecular Biotechnology*, Vol. 74, pp. 277-302.

Nahshol, O., Bronner, V., Notcovich, A., Rubrecht, L., Laune, D. & Bravman, T. (2008). Parallel kinetic analysis and affinity determination of hundreds of monoclonal antibodies using the ProteOn XPR36. *Analytical Biochemistry*, Vol. 383, No. 1, pp. 52-60.

Neuberger, M. S. (2008). Antibody diversity by somatic mutation: from Burnet onwards. *Immunology Cell Biol*, Vol. 86, No. 2, pp. 124-132.

Nissim, A., Hoogenboom, H. R., Tomlinson, I. M., Flynn, G., Midgley, C., Lane, D. & Winter, G. (1994). Antibody fragments from a 'single pot' phage display library as immunochemical reagents. *Embo Journal*, Vol. 13, No. 3, pp. 692-698.

Osbourn, J. K., Field, A., Wilton, J., Derbyshire, E., Earnshaw, J. C., Jones, P. T., Allen, D. & McCafferty, J. (1996). Generation of a panel of related human scFv antibodies with high affinities for human CEA. *Immunotechnology*, Vol. 2, No. 3, pp. 181-196.

Pasqualini, R. & Ruoslahti, E. (1996). Organ targeting in vivo using phage display peptide libraries. *Nature*, Vol. 380, No. 6572, pp. 364-346.

Pedersen, H., Hölder, S., Sutherlin, D. P., Schwitter, U., King, D. S. & Schultz, P. G. (1998). A method for directed evolution and functional cloning of enzymes. *Proceedings of the National Academy of Sciences USA*, Vol. 95, No. 18, pp. 10523-10528.

Rajpal, A., Beyaz, N., Haber, L., Cappuccilli, G., Yee, H., Bhatt, R. R., Takeuchi, T., Lerner, R. A. & Crea, R. (2005). A general method for greatly improving the affinity of antibodies by using combinatorial libraries. *Proceedings of the National Academy of Sciences USA*, Vol. 102, No. 24, pp. 8466-8471.

Razai, A., Garcia-Rodriguez, C., Lou, J., Geren, I. N., Forsyth, C. M., Robles, Y., Tsai, R., Smith, T. J., Smith, L. A., Siegel, R. W., Feldhaus, M. & Marks, J. D. (2005). Molecular evolution of antibody affinity for sensitive detection of botulinum neurotoxin type A. *Journal of Molecular Biology*, Vol. 351, No. 1, pp. 158-169.

Safsten, P., Klakamp, S. L., Drake, A. W., Karlsson, R. & Myszka, D. G. (2006). Screening antibody-antigen interactions in parallel using Biacore A100. *Analytical Biochemistry*, Vol. 353, No. 2, pp. 181-190.

Sanna, P. P., Williamson, R. A., De Logu, A., Bloom, F. E. & Burton, D. R. (1995). Directed selection of recombinant human monoclonal antibodies to herpes simplex virus glycoproteins from phage display libraries. *Proceedings of the National Academy of Sciences USA*, Vol. 92, No. 14, pp. 6439-6443.

Schaffitzel, C., Berger, I., Postberg, J., Hanes, J., Lipps, H. J. & Pluckthun, A. (2001). In vitro generated antibodies specific for telomeric guanine-quadruplex DNA react with Stylonychia lemnae macronuclei. *Proceedings of the National Academy of Sciences USA*, Vol. 98, No. 15, pp. 8572-8577.

Schier, R., Bye, J., Apell, G., McCall, A., Adams, G. P., Malmqvist, M., Weiner, L. M. & Marks, J. D. (1996). Isolation of high-affinity monomeric human anti-c-erbB-2 single chain Fv using affinity-driven selection. *Journal of Molecular Biology*, Vol. 255, No. 1, pp. 28-43.

Schreuder, M. P., Brekelmans, S., van den Ende, H. & Klis, F. M. (1993). Targeting of a heterologous protein to the cell wall of Saccharomyces cerevisiae. *Yeast*, Vol. 9, No. 4, pp. 399-409.

Schreuder, M. P., Mooren, A. T., Toschka, H. Y., Verrips, C. T. & Klis, F. M. (1996). Immobilizing proteins on the surface of yeast cells. *Trends Biotechnology*, Vol. 14, No. 4, pp. 115-120.

Seelig, B. & Szostak, J. W. (2007). Selection and evolution of enzymes from a partially randomized non-catalytic scaffold. *Nature*, Vol. 448, No. 7155, pp. 828-831.

Sheets, M. D., Amersdorfer, P., Finnern, R., Sargent, P., Lindquist, E., Schier, R., Hemingsen, G., Wong, C., Gerhart, J. C. & Marks, J. D. (1998). Efficient construction of a large nonimmune phage antibody library: the production of high-affinity human single-chain antibodies to protein antigens. *Proceedings of the National Academy of Sciences USA*, Vol. 95, No. 11, pp. 6157-6162.

Smith, G. P. (1985). Filamentous fusion phage: novel expression vectors that display cloned antigens on the virion surface. *Science*, Vol. 228, No. 4705, pp. 1315-1317.

Soderlind, E., Strandberg, L., Jirholt, P., Kobayashi, N., Alexeiva, V., Aberg, A. M., Nilsson, A., Jansson, B., Ohlin, M., Wingren, C., Danielsson, L., Carlsson, R. & Borrebaeck, C. A. (2000). Recombining germline-derived CDR sequences for creating diverse single-framework antibody libraries. *Nature Biotechnology*, Vol. 18, No. 8, pp. 852-856.

Soumillion, P. & Fastrez, J. (2001). Novel concepts for selection of catalytic activity. *Current Opinion in Biotechnology*, Vol. 12, No. 4, pp. 387-394.

Soumillion, P., Jespers, L., Bouchet, M., Marchand-Brynaert, J., Winter, G. & Fastrez, J. (1994). Selection of beta-lactamase on filamentous bacteriophage by catalytic activity. *Journal of Molecular Biology*, Vol. 237, No. 4, pp. 415-422.

Stijlemans, B., Conrath, K., Cortez-Retamozo, V., Van Xong, H., Wyns, L., Senter, P., Revets, H., De Baetselier, P., Muyldermans, S. & Magez, S. (2004). Efficient targeting of conserved cryptic epitopes of infectious agents by single domain antibodies. African trypanosomes as paradigm. *Journal of Biological Chemistry*, Vol. 279, No. 2, pp. 1256-1261.

Thom, G., Cockroft, A. C., Buchanan, A. G., Candotti, C. J., Cohen, E. S., Lowne, D., Monk, P., Shorrock-Hart, C. P., Jermutus, L. & Minter, R. R. (2006). Probing a protein-protein interaction by in vitro evolution. *Proceedings of the National Academy of Sciences USA*, Vol. 103, No. 20, pp. 7619-7624.

Tordsson, J., Abrahmsen, L., Kalland, T., Ljung, C., Ingvar, C. & Brodin, T. (1997). Efficient selection of scFv antibody phage by adsorption to in situ expressed antigens in tissue sections. *Journal of Immunological Methods*, Vol. 210, No. 1, pp. 11-23.

Ulrich, H. D., Patten, P. A., Yang, P. L., Romesberg, F. E. & Schultz, P. G. (1995). Expression studies of catalytic antibodies. *Proceedings of the National Academy of Sciences USA*, Vol. 92, No. 25, pp. 11907-11911.

van den Beucken, T., Pieters, H., Steukers, M., van der Vaart, M., Ladner, R. C., Hoogenboom, H. R. & Hufton, S. E. (2003). Affinity maturation of Fab antibody fragments by fluorescent-activated cell sorting of yeast-displayed libraries. *FEBS Letters*, Vol. 546, No. 2-3, pp. 288-294.

VanAntwerp, J. J. & Wittrup, K. D. (2000). Fine affinity discrimination by yeast surface display and flow cytometry. *Biotechnology Prog*, Vol. 16, No. 1, pp. 31-37.

Vaughan, T. J., Williams, A. J., Pritchard, K., Osbourn, J. K., Pope, A. R., Earnshaw, J. C., McCafferty, J., Hodits, R. A., Wilton, J. & Johnson, K. S. (1996). Human Antibodies with sub-nanomolar affinities isolated from a large non-immunized phage display library. *Nature Biotechnology*, Vol. 14, No. 3, pp. 309-314.

Wassaf, D., Kuang, G., Kopacz, K., Wu, Q. L., Nguyen, Q., Toews, M., Cosic, J., Jacques, J., Wiltshire, S., Lambert, J., Pazmany, C. C., Hogan, S., Ladner, R. C., Nixon, A. E. & Sexton, D. J. (2006). High-throughput affinity ranking of antibodies using surface plasmon resonance microarrays. *Analytical Biochemistry*, Vol. 351, No. 2, pp. 241-253.

Yang, W. P., Green, K., Pinz-Sweeney, S., Briones, A. T., Burton, D. R. & Barbas, C. F. r. (1995). CDR walking mutagenesis for the affinity maturation of a potent human anti-HIV-1 antibody into the picomolar range. *Journal of Molecular Biology*, Vol. 254, No. 3, pp. 392-403.

Yelton, D. E., Rosok, M. J., Cruz, G., Cosand, W. L., Bajorath, J., Hellstrom, I., Hellstrom, K. E., Huse, W. D. & Glaser, S. M. (1995). Affinity maturation of the BR96 anti-carcinoma antibody by codon-based mutagenesis. *Journal of Immunology*, Vol. 155, No. 4, pp. 1994-2004.

Zahnd, C., Amstutz, P. & Pluckthun, A. (2007). Ribosome display: selecting and evolving proteins in vitro that specifically bind to a target. *Nature Methods*, Vol. 4, No. 3, pp. 269-279.

Zahnd, C., Spinelli, S., Luginbuhl, B., Amstutz, P., Cambillau, C. & Pluckthun, A. (2004). Directed in vitro evolution and crystallographic analysis of a peptide-binding single chain antibody fragment (scFv) with low picomolar affinity. *Journal of Biological Chemistry*, Vol. 279, No. 18, pp. 18870-18877.

5

Strengths and Weaknesses of Selected Modeling Methods Used in Systems Biology

Pascal Kahlem et al.*
EMBL -European Bioinformatics Institute,
Wellcome Trust Genome Campus, Hinxton,
Cambridge, United Kingdom

1. Introduction

The development of multicellular organisms requires the coordinated accomplishment of many molecular and cellular processes, like cell division and differentiation, as well as metabolism. Regulation of those processes must be very reliable and capable of resisting fluctuations of the internal and external environments. Without such homeostatic capacity, the viability of the organism would be compromised. Cellular processes are finely controlled by a number of regulatory molecules. In particular, transcription factors are amongst the proteins that determine the transcription rate of genes, including those involved in development and morphogenesis. For this reason, the molecular mechanisms responsible for the homeostatic capacity and coordinated behavior of the transcriptional machinery have become the focus of several laboratories.

Modern techniques of molecular genetics have greatly increased the rate at which genes are recognized and their primary sequences determined. High throughput methods for the identification of gene-gene and protein-protein interactions exist and continue to be refined, raising the prospect of high-throughput determination of networks of interactions in discrete cell types. Still, classic biochemical and physiological studies are necessary to identify the targets, and to understand the functions of the encoded proteins. For such reasons, the rate at which pathways are described is much slower than the rate of genome sequencing. The large quantity of available sequences creates the challenge for molecular geneticists of linking genes and proteins into functional pathways or networks. Biologists are often interested in particular subsets of these very extensive networks obtained with

---

* Alessandro DiCara[2], Maxime Durot[3], John M. Hancock[4], Edda Klipp[5], Vincent Schächter[3], Eran Segal[7], Ioannis Xenarios[8], Ewan Birney[1] and Luis Mendoza[6].
[1]EMBL -European Bioinformatics Institute, Wellcome Trust Genome Campus, Hinxton, Cambridge, United Kingdom
[2] Merck Serono, 9, chemin des Mines, Geneva, Switzerland
[3] UMR 8030 CNRS, Université d'Evry, CEA/Genoscope– CEA, 2Evry, France
[4]MRC Mammalian Genetics Unit, Harwell, Oxfordshire, United Kingdom
[5]Theoretical Biophysics, Institute for Biology, Humboldt University Berlin, Berlin, Germany
[6]Instituto de Investigaciones Biomédicas, Universidad Nacional Autónoma de México, Ciudad Universitaria, México.
[7]Department of Computer Science and Applied Mathematics, Weizmann Institute of Science, Rehovot, Israel
[8]Swiss Institute of Bioinformatic, Quartier Sorge - Batiment Genopode, Lausanne Vaud, Switzerland

high-throughput techniques, subsets that are involved in accomplishing some specific biological objective. Such a view often does not take into account the global dynamical properties of networks, which may be finally necessary to understand the behavior of the system, for instance to correlate genotype and phenotype. This is where computational models of some particular networks will support bench biologists by providing their descriptive and predictive capacity.

With the constant development of faster and more reliable biotechnologies, the scientific community is presented with a growing collection of biological information, some qualitative, some quantitative. Formal databases include catalogs of genes (EnsEMBL (Hubbard et al. 2007)), proteins (UniProt (The UniProt Consortium 2007)), enzymes and their substrates (BRENDA (Chang et al. 2009)) and molecular reactions (Reactome (Matthews et al. 2009)), many from multiple species. Quantitative data resulting from large-scale experiments are also collected in databases; some of them are public, including gene expression (ArrayExpress (Parkinson et al. 2007)), protein interactions (IntAct (Kerrien et al. 2007)), reaction kinetics (SABIO-RK (Rojas et al. 2007)), cellular phenotypes (MitoCheck (Erfle et al. 2007)) and whole organism phenotypes (e.g. EuroPhenome (Morgan et al. 2010)) amongst others. The combination of these biological data with mathematical probabilistic methods to produce models for selected domains of biological systems becomes essential to better understand and possibly predict the behavior of systems in conditions that cannot be experimentally assayed.

However, the diversity of biological systems, the system-specific type of information and the limited availability of data, implies a requirement for the development of adapted modeling methods. Modeling methods must be tailored not only to make the best use of the information available but also to answer specific biological questions, which can span from the understanding of the function of a pathway and its evolution to the global molecular mechanisms underlying cellular events such as cell differentiation for example.

Figure 1 illustrates the tradeoff between network size and the possibility to model different aspects – from topology to dynamics – of the network. We will give here an overview of the current progress of modeling methods illustrated by chosen examples of systems.

2. Sources of data

Data acquisition has been greatly improved by the application of high-throughput screening methods in molecular biology, enabling simultaneous measurement of thousands of molecular events (data points) in a single experiment. Besides, public repositories for high-content biological data (ArrayExpress, IntAct, MitoCheck, amongst others) enable, with the increasing use of standardized annotations, data integration at the system level (Table 1).

The choice of the type of data needed to reconstruct networks depends on the type and size of system under study and is driven by the scientific rationale. Although all sources of experimental data are potentially useful for modeling a system, modeling specifically a signaling pathway will have different requirements in terms of data types to be integrated than a metabolic pathway. Additionally, the granularity of models' description can also be adjusted to incorporate the extent of experimental data available.

The TGF-beta signaling pathway has been extensively studied by the authors using both kinetic and Boolean approaches. Modeling the TGF-beta signaling pathway requires information to build the topology of the network, from recognition of the signal to its transduction to the nucleus, including the down-stream effect on the transcriptome. This information will be obtained from methods of molecular biology and biochemistry. Studying protein interactions will be necessary for predicting the topology of the model, while assessing protein amounts and phosphorylation states, genes' expression levels, and reaction kinetics will be necessary for recording the dynamics of the model, thereby not only reducing the number of parameters to be predicted, but also defining constraints in the network. Modeling a metabolic pathway will have other data requirements, such as classical biochemistry and genetics to determine the network topology, metabolomics to inform flux parameters and, if the system allows it, phenotypes may also be integrated into the model.

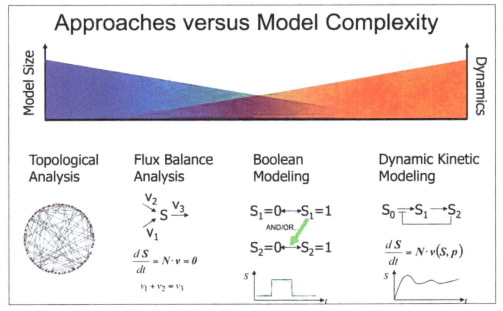

Fig. 1. The size of the network and the available amount of data determine the type of approach to tackle biological questions. For very large networks, exploration is often restricted to topological aspects. Metabolic networks, even of large size can be studied with flux balance analysis, given that stoichiometry is known. For medium size networks, application of Boolean logic allows to detect different qualitative modes of dynamics. Kinetic modeling is mainly restricted to small networks that are already well characterized. Advances in data acquisition and modeling techniques will extend the applicability of different analysis methods to various biological processes in the foreseeable future.

3. Procedures of network reconstruction

Determining the network topology defines the limits of the system under study and enables the incorporation of stoichiometric information and experimental parameters (see Table 1).

Over the last decades, systems' modeling has been successfully applied to formulate biological problems in a mathematically tractable form. A number of general approaches and specific models were developed to enable the understanding of biological phenomena: For example, the introduction of metabolic control analysis (Heinrich and Rapoport 1974; Kacser and Burns 1973) pinpointed the fact neglected earlier that metabolic regulation is a property of both the structure of the metabolic network and the kinetics of the individual

Table 1. Examples of sources of information (databases) related with models and methods.

	Type of data	Name of the database	URL	Modeling methods		
				ODE	Boolean	Constraint-based
Knowledge Databases	Small chemical entities	ChEBI	www.ebi.ac.uk/chebi			
	Gene Ontology	GO	www.geneontology.org		x	
	Taxonomy	Uniprot taxonomy	www.uniprot.org/taxonomy		x (localization)	
	Genetic diseases	OMIM	www.ncbi.nlm.nih.gov/omim			
	Literature	CiteXplore	www.ebi.ac.uk/citexplore		x	
	Genomes	Ensembl	www.ensembl.org			
	Nucleotide Sequences	EMBL-Bank	www.ebi.ac.uk/embl			
	Proteins	UniProt	www.uniprot.org		x	
	Enzymes	IntEnz	www.ebi.ac.uk/intenz	x	x	
	Structure	MSD	www.ebi.ac.uk/msd			
	Enzymes	BRENDA	www.brenda-enzymes.info	x		
	Pathways	Reactome	www.reactome.org	x	x	
	Pathways	KEGG	www.genome.jp/kegg	x	x	
	Pathways	Chillibot	www.chillibot.net		x	
	Pathways	iHOP	www.ihop-net.org/UniPub/iHOP		x	
	Pathways	Panther	www.pantherdb.org/panther		x	
	Pathways	Ingenuity	www.ingenuity.com		x	
	Pathways	GeneGO	www.genego.com		x	
	Pathways	BioCyc	www.biocyc.org		x	
	Genome-scale metabolic models	Thermodynamics SEED models	www.theseed.org/models			x
	Biological Models	BioModels	www.ebi.ac.uk/biomodels	x		x
	Transporters	TransportDB	www.membranetransport.org			x
	Thermodynamics	NIST	xpdb.nist.gov/enzyme_thermodynamics/	x		
Quantitative Databases	Gene expression	Array Express	www.ebi.ac.uk/aerep			
	Protein interaction	IntAct	www.ebi.ac.uk/intact			
	Reaction Kinetics	SABIO-RK	sabiork.villa-bosch.de	x		
	Cellular phenotypes	MitoCheck	www.mitocheck.org			
	Growth phenotypes	ASAP	asap.ahabs.wisc.edu/asap/home.php			x
	Metabolite concentrations	Escherichia coli Multi-omics Database	ecoli.iab.keio.ac.jp			x
	Reaction fluxes	Escherichia coli Multi-omics Database	ecoli.iab.keio.ac.jp			x
	Kinetic Parameters	KMedDB	sysbio.molgen.mpg.de/KMedDB	x		

enzymes, instead of only the task of a single rate-limiting enzyme. Detailed metabolic modeling has highlighted among other aspects that knowledge of *in vitro* kinetics of isolated enzymes is often not sufficient to understand metabolism (Teusink et al. 2000). The study of signaling pathways showed how cells process information and has revealed many regulatory motifs including negative or positive feedback and bistability (for overviews see e.g. (Ferrell 2002; Tyson, Chen, and Novak 2003)). Integration of signaling with gene expression, metabolism and biophysical changes demonstrated the contribution of various components of the cellular networks to stress response (e.g. (Klipp et al. 2005)). A number of problems have been tackled with modelling: Among other examples, we find

i. The question of explaining experimentally observed oscillations (e.g. in metabolism). This work resulted in various oscillator models such as the Higgins-Sel'kov oscillator (Higgins 1964; Sel'kov 1968) and more complicated models (e.g. (Hynne, Dano, and Sorensen 2001)) or application to interacting yeast cells (Wolf and Heinrich 2000; Wolf et al. 2000);
ii. Can we understand cell cycle progression from protein interactions? The approaches vary from very simple models (Goldbeter 1991) to comprehensive studies (e.g. (Chen et al. 2004));
iii. What determines robustness of bacterial chemotaxis? Barkai and Leibler proposed a mechanism for robust adaptation in simple signal transduction networks (Barkai and Leibler 1997).

Many of these examples seek to explain emergent properties whose origins are not obvious from a cursory examination of the underlying interactions. In addition, such models helped to establish an abstract language for describing biological observations and to introduce concepts such as equilibrium or steady state (early) or control, stability, robustness or signal amplification (later) into analysis of biological systems.

3.1 Types of networks

Biological networks reflect the regulatory and functional interactions between molecular components (genes, proteins, metabolites) but may also be extended to integrate information on cellular behavior and physiological impact.

The type of information usually represented in networks can be heterogeneous. For instance, the transfer of information through regulatory interactions can be distinguished from the transfer of mass during metabolic reactions.

Computational models are constrained by the amount of information available in the system of interest, which is itself limited to the fabrication of biotechnological tools enabling scientific explorations of various biological systems at the molecular scale. To date, knowledge on prokaryote metabolism is rather complete in comparison to eukaryotic organisms. Similarly, while biological data can be extracted using monocellular cultures in vitro, the transposition of knowledge to multicellular systems remains uncertain.

3.2 Modeling methods
3.2.1 ODE modeling

Among the most frequently applied techniques for modeling dynamic processes in biological networks are systems of ordinary differential equations (ODE) (Klipp 2007). These systems are used to describe the temporal changes of molecular concentrations caused by production, degradation, transport, or modification of the modeled substances. Such changes of concentration are expressed as a function of rates of reaction and appropriate stoichiometric coefficients. Reaction rates, in turn, can be of several types, such as the mass action law (Guldberg and Waage 1879), the Michaelis-Menten rate law (Briggs and Haldane 1925; Michaelis and Menten 1913), or more complicated forms to attain some specific kinetics (Cornish-Bowden et al. 2004; Klipp et al. 2005; Koshland, Nemethy, and Filmer 1966; Liebermeister and Klipp 2006; Monod, Wyman, and Changeux 1965).

The use of kinetic ODE modeling using rate laws has been the cornerstone of our traditional biochemical terminology and thinking. Such equations have proven quite successful providing a convenient language and conveying immediate meaning since the formulation of the mass action law (Waage and Guldberg 1864), and the modeling of enzyme kinetics (Michaelis and Menten 1913). As a consequence, there is a vast amount of research and huge numbers of publications that have been devoted to the modeling of biochemical reactions using ordinary differential equations (Heinrich and Rapoport 1974; Tyson, Chen, and Novak 2003). A widely successful example of the use of the kinetic approach is the modeling of the cell-cycle control in yeast (Tyson, Csikasz-Nagy, and Novak 2002).

A severe drawback of the ODE approach is the large number of parameters involved in the system of equations. This implies that for any given biological system, there is the need of large sets of experimental data to determine the parameter values of the equations. Moreover, although a number of kinetic parameters are already available in databases, these values are not always applicable for other organisms or other experimental scenarios than for those for which they were measured. For example, kinetic constants usually are developed for metabolic reactions catalyzed by enzymes acting in a test tube, so the appropriateness for modeling in vivo reactions remains to be proven. They may also have been measured in different species from the one under consideration. Alternatively, there is the possibility of parameters estimation (Ashyraliyev, Jaeger, and Blom 2008; Sorribas and Cascante 1994), but such methodology is computationally costly and there is no guarantee that the computational result is biologically correct.

Most ODE models make use of continuous functions to describe the kinetic laws. Continuous functions, however, may not always be appropriate for describing biological processes. For example, given that molecules are discrete entities, the number of molecules as a function in time is in reality a discrete function. Hence, it is important to assess if the use of a continuous function in a given model is a reasonable approximation to reality. As a rule of thumb, if the experimental error at measuring the real value of a variable is larger than the jump in the discrete value, then it is usually harmless to replacement discrete functions by continuous ones.

The representation of chemical species as a concentration with the use of continuous variables also assumes that the system is absolutely uniform. However, in reality biochemical systems frequently exhibit a large degree of spatial heterogeneity due to processes such as compartmentalization, molecular association, or restricted diffusion. It has been mathematically demonstrated that changes in the spatial distribution of molecules have a large impact on the dynamical behavior of a biochemical system (Zimmerman and Minton 1993).

Often ODE models are deterministic, but under certain circumstances, a deterministic approach does not give an adequate representation of the biological system. At the molecular scale, individual molecules randomly collide with one another, allowing for a chemical reaction to occur only if the collision energy is strong enough. This effect is observed for small volumes; when using deterministic equations the smaller the volume the less accurate the model becomes (Ellis 2001; Erdi and Toth 1989). Therefore, it is important to make sure that the system under study is large enough to avoid stochastic fluctuations (Fournier et al. 2007) which could be amplified and thus originate observable macroscopic effects.

3.2.2 Discrete networks

A network is a system formed by multiple nodes, each associated with a state of activation, which depends upon the state of a set of nodes. A common practice is to represent nodes with the use of continuous variables. However, usually there is only qualitative experimental data regarding the activity of most genes and/or proteins. For such reasons, a useful simplification is to suppose that genes can attain only a finite number of possible states, thus allowing their representation with discrete variables. In the simplest case, a gene might be "turned on" (or "active", or "1") or "turned off" (or "inactive", or "0") at a given time. In this case, we are dealing with binary networks, also known as Boolean networks. Boolean networks were first presented by Kauffman (Glass and Kauffman 1973; Kauffman 1969) so as to give a qualitative description of the concerted action of a group of genes during cellular differentiation. Such models were originally developed as a suitable simplification for the analysis of genetic regulation, and were originally studied exclusively from a statistical point of view (Kauffman 1993) due to the lack of biological data on experimentally validated biological networks. More recently, Boolean network models have been developed for a series of biological systems showing the suitability of this methodology to capture key aspects of cellular development and differentiation (Albert and Othmer 2003; Davidich and Bornholdt 2008; Faure et al. 2006; Gupta et al. 2007; Huang and Ingber 2000; Kervizic and Corcos 2008; Li et al. 2004; Mendoza, Thieffry, and Alvarez-Buylla 1999; Saez-Rodriguez et al. 2007; Samal and Jain 2008).

The use of Boolean models has permitted the discovery of the influence of the network topology on its dynamical behavior. Specifically, most studied networks include feedback loops or circuits. It has been shown that their presence is necessary to ensure multistationarity and homeostasis, which are particularly important properties of biological systems. The logical analysis of feedback loops decomposes any network into a well-defined set of feedback loops. It was first developed by Thomas (Thomas 1973), and formally demonstrated by others (Gouzé 1998; Plahte, Mestl, and Omholt 1995; Snoussi 1998). Negative feedback loops generate homeostasis in the form of damped or sustained oscillations. The importance of homeostasis in maintaining the internal environment of an organism is well known and dates from the work of (Cannon 1929). Conversely, positive feedback loops generate multiple alternative steady states or multistationarity. The biological interpretation of multistationarity as cellular differentiation goes back to Delbrück (Delbrück 1949), and has been developed by the group of Thomas (Thieffry et al. 1995; Thomas 1973).

Boolean networks are widely used for their computational tractability and their capability of providing qualitatively correct results; there is, however, an important issue to take into account. Boolean networks describe time as a discrete variable, hence there is the need to decide at each clock tick which nodes of the network are going to be updated. On the one hand, in the synchronous approach all nodes are updated at each time step. This methodology is the easiest to implement but also the less realistic, since it is highly unlikely that all molecules in the modeled network have the same time response. On the other hand, in the asynchronous approach only one node is updated at each time step. While this approach is closer to reality, there is usually no experimental information regarding the correct order of response. Worse still, the group of stable states attained by the system using the synchronous and asynchronous approaches are not necessarily identical. It is therefore advisable to use both methodologies, with the aid of a modeling software (Garg, Banerjee, and De Micheli 2008; Gonzalez et al. 2006), and then use biological knowledge to decide among all possible outcomes.

3.2.3 Qualitative modeling

The use of continuous variables provides fine-granularity modeling, thus allowing for the description of a richer dynamical behavior. However, experimental data to support parameter fitting is very scarce; hence the development of quantitative models of regulatory networks is limited to a small set of experimental systems for which a large quantity of molecular data has been gathered (for an example see (Jaeger et al. 2004)). The alternative approach of modeling with the use of Boolean networks is not always possible, though, usually because of the lack of experimental information to infer the logical rules governing the response of nodes. There are, however, intermediate modeling methodologies lying between the coarse-grained binary approach and the fine-grained use of ordinary-differential equations: among them it is possible to find the use of piecewise-linear differential equations, qualitative differential equations and standardized qualitative dynamical systems.

Piecewise-linear models have been proposed for the modeling of regulatory networks. These equations, originally proposed in (Glass and Kauffman 1973) have been amply studied (Glass and Pasternack 1978; Gouzé and Sari 2002; Mestl, Plahte, and Omholt 1995; Plahte, Mestl, and Omholt 1994) from a theoretical point of view. Variables in the piecewise-linear approach represent the concentrations of proteins, while the differential equations describe the regulatory interactions among genes encoding these proteins. Each differential equation contains two terms, namely the activation part consisting of a weighted sum of products of step functions and the decay rate. The mathematical form of these equations divides the state space into multidimensional boxes, and inside the volume of each box the equations are reduced to linear ODEs, making the behavior of the system inside a given volume straightforward to analyze. Nevertheless, the global behavior of these systems of equations can be very complex, with chaotic behavior being rather common (Lewis and Glass 1995; Mestl, Bagley, and Glass 1997). Despite this drawback piecewise-linear differential equations have been used to analyze several regulatory networks of biological interest (De Jong et al. 2004; Ropers et al. 2006; Viretta and Fussenegger 2004).

In the case of the use of qualitative differential equations, the dependent variable takes a qualitative value composed of a qualitative magnitude and a direction. Here, the qualitative magnitude is a discretization of a continuous variable, while the qualitative direction is the sign of its derivative. Furthermore, each equation is actually a set of constraints that restrict the possible qualitative values of the variable. To solve the system, it is necessary to create a tree of possible sequences of transitions from the initial state. Now, this characteristic makes the methodology difficult to apply for large biological systems, since the trees describing the dynamical behavior rapidly grow out of bounds. This scalability problem has restricted the application of qualitative equations to a small number of models (Heidtke and Schulze-Kremer 1998; Trelease, Henderson, and Park 1999).

While Boolean models approximate a continuous sigmoid by a discontinuous step function, the standardized qualitative dynamical systems method goes the other way around. Within this methodology, the network is modeled as a continuous system using a set of ordinary differential equations. These equations describe the rate of change of activation (or synthesis, or transcription) as a sigmoid function of the state of activation of the controlling input variables. The variables representing the state of activation are normalized, so that they are constrained in the range [0,1]. This feature enables a comparison of the dynamics against the results of a purely binary model. The characteristic that distinguishes this method from other approaches is that the equations are normalized and standardized, i.e. they are not network-specific. Moreover, models can be developed even in the total absence of information regarding the molecular mechanisms of a given network because no stoichiometric or rate constants are needed. Given that this qualitative method was recently developed, it has been used to model a small number of biological networks (Mendoza and Pardo 2010; Mendoza and Xenarios 2006; Sanchez-Corrales, Alvarez-Buylla, and Mendoza 2010). The standardized qualitative dynamical systems methodology has been used for the development of a software package for the automated analysis of signaling networks (http://www.enfin.org/squad).

3.2.4 Constraint-based modeling

Analyzing the dynamics of large-scale metabolic networks using kinetic modeling techniques is hampered by the size of the biological system because of the large number of parameters that need to be fitted. Constraint-based modeling bridges the gap between the mere static representation of graph-based methods and the detailed dynamic pathway analyses of kinetic modeling techniques (Feist and Palsson 2008). This framework aims at simulating at the cellular level the global dynamics of metabolism, using a limited amount of information. To that end, the method is based on the description of all reaction fluxes that are compatible with constraints deriving from basic physical assumptions, specific biological information or experimental measures. The assumption of steady state dynamics, for example, simplifies the analysis by only requiring knowledge on the reaction stoichiometries, which can be easily obtained from metabolic databases (Kanehisa et al. 2006; Karp, Paley, and Romero 2002; Matthews et al. 2009). Other information such as reaction reversibility (Joyce and Palsson 2006), flux measurements, gene expression (Shlomi et al. 2008) and metabolite concentration (Henry et al. 2006; Kummel, Panke, and Heinemann 2006) can be incorporated into the modeling if available (Durot, Bourguignon, and Schachter 2009). Compared to kinetic models, constraint-based models are therefore easily reconstructed at large scale (Feist et al. 2009) and online resources that provide automatically reconstructed models are currently being launched (Henry et al. 2010).

A set of constraints narrows the total number of possible flux states, which represent the set of possible metabolic behaviors of the system. In order to explore this set of attainable flux distributions, a number of mathematical and computational methods have been designed (Durot, Bourguignon, and Schachter 2009). Some of them attempt to describe the overall set of possibilities either by enumerating all basic independent metabolic routes, known as elementary modes or extreme pathways (Papin et al. 2004), or by randomly sampling the space of flux distributions (Schellenberger and Palsson 2009). Other methods focus on specified metabolic objectives, e.g. the production of biomass components, and look for flux distributions that optimize it (Varma and Palsson 1994).

The constraint-based modeling framework utilizes a set of equations describing metabolites, where each equation contains as many unknowns as there are fluxes for a given metabolite. Very often the number of unknowns is larger than the number of equations. This is the main drawback of the approach, since it means that the system of equations has an infinite number of solutions. This problem is usually solved by maximizing some functions, which in turn assumes that the modeled organism has evolved optimal metabolic pathways, which is not necessarily true. Despite this limitation, the constraint-based modeling approach has been successfully applied for predicting growth phenotypes for knock-out mutants on varying environments, analyzing essentiality or dispensability, integrating genome-scale experimental data, and driving metabolic engineering designs (Durot, Bourguignon, and Schachter 2009; Feist and Palsson 2008).

3.2.5 Probabilistic graphical models

Probabilistic graphical models (Pearl 1988) represent another class of methods that have recently gained much popularity in studies of gene regulation. Such models always define a joint probability distribution over all the properties in the domain, where a property is represented by a random variable, which is either hidden (e.g., the functional module that a gene belongs to) or observed (e.g., the expression level of a gene). Random variables and the probabilistic dependencies among them define a probabilistic graphical model, which provides a statistical framework for representing complex interactions in biological domains in a compact and efficient manner.

In contrast to procedural methods, such as data clustering followed by motif finding (Tavazoie et al. 1999), probabilistic graphical models are, as their name implies, declarative, model-based approaches, where by a model we mean a simplified description of the biological process that could have generated the observed data. An important property of these models is their ability to handle uncertainty in a principled way, which is particularly useful in the biological domain, due to the stochastic nature of the biological system and due to the noise in the technology for measuring its properties. The details of the models (dependency structure and parameters) are typically learned automatically from the data, where the goal is to find a model that maximizes the probability of the model given the observed data. The main difficulty with this approach is that the learning task is challenging, as it involves several steps, some of which are computationally intractable.

Despite the computational difficulties of the probabilistic approach, it has been successfully used in several cases. One example is the reconstruction of the structure of regulatory networks (Friedman et al. 2000). The main idea is that if the expression of gene A is regulated by proteins B and C, then A's expression level is a function of the joint activity levels of B and C; this can easily be extended to networks composed of modules of co-regulated genes (Gasch et al. 2000; Segal et al. 2003). Another example is the case where probabilistic graphical models were applied to identify the cis-regulatory motifs through which transcriptional programs occur (Segal, Yelensky, and Koller 2003), with the aim of identifying modules of co-regulated genes and explain the observed expression patterns of each module via a motif profile that is common to genes in the same module.

4. Strengths and weaknesses of modeling methods

The methods chosen to model a given system depend mostly on the qualitative and quantitative features of biological data available to construct the model and on the biological question that is asked. To be accurate, an ODE model will require as much experimental information as possible since all parameters of the network are represented and missing parameters will be estimated to fit experimental values. Boolean modeling minimizes the network to the essential nodes that are subjected to decisional events and feedback loops. Although such a reductionist method does not involve kinetic parameters, it remains capable of predicting the logical behavior of the network and identifying steady states. Although these two previous methods work with currently size-limited networks, global metabolic modeling using the constraint-based method allows the comparative analysis of thermodynamic fluxes between chosen states of a system. Probabilistic graphical models integrate different types of biological information, such as gene expression data for example, in order to reconstruct the networks and predict their behaviors.

The models studied by the authors are developed in synergy by wet and dry laboratories, which cycle between computational modeling and experimental testing. For example, the human TGF-beta signaling pathway has been modeled using ODE modeling by the group of E. Klipp (Zi and Klipp 2007) and using Boolean modeling by the group of I. Xenarios. The ODE modeling required a large amount of biochemical information, or used parameter estimation methods to predict missing values. However, since this method is based on real biological data, it allowed investigating precisely early time points of the TGF-beta signaling cascade. The Boolean method required obviously less biochemical information than the ODE, which allowed to compute accurately a large network articulated across several positive or negative feedback loops. However, qualitative models despite looking fairly simple to construct pose several issues when one wants to validate experimentally the proposed predictions. Each component (node) of a model is represented as an abstraction of a molecular entity. For example, the interferon gamma receptor complex will be represented in qualitative models as a single entity (node). However, from a biochemical point of view interferon gamma receptor is composed of three subunits (a,b,g). This raises the question whether all the subunits should be experimentally measured or only one of them. This situation was experienced by the authors who modelled the TGFb pathway: Several predictions were produced after reconstructing the Boolean network. One of them was that knocking-out one component introduced an oscillation of several components of the TGF-beta pathway. However, upon experimental validation, the nodes that were suggested to oscillate did not show any changes at the mRNA/protein expression or phosphorylation states during a time-course experiment. The model did not indicate at what molecular level the predicted oscillation would be measurable.

According to the experience acquired with the models developed by the authors as part of the ENFIN Consortium, we have assessed the relationship between modeling frameworks and the biological questions which could be answered in each case (Table 2).

Besides the biological question and the subjective choice of the modeler, the ability to access experimental data can be critical for the choice of a modeling method. This is schematically represented in the decision tree (Figure 2).

Testing entire model predictions with wet experiments is often impossible, because of simple limitations of available technologies. Experimental assays are therefore designed to target key components of models, a strategy often taken in industry to prioritize pharmaceutical targets. To our experience, the integration of several disciplines improves the coverage of computational analysis. Improving computational models can also require repetition of experiments potentially needing expensive settings, but which may not result in genuine scientific discovery. For instance, there is no point measuring some parameters that may be useful for the model if the accuracy of measurement is too poor to resolve between alternative model versions. Problems may also arise concerning the spatial resolution of experimental data. For example, how useful are whole animal metabolomics in metazoans for assessing the behavior of a particular cell type?

Biological Question	Boolean networks or qualitative models (of signaling or regulatory networks)	Probabilistic graphical models of regulatory networks	Constraint-based models (purely metabolic or complemented by a Boolean regulatory layer)	Dynamic ODE models
Reconstruct the interaction structure of the cellular process from experimental data	** using expression data, protein interaction data, perturbation experiments (see III.2.d) with expression data optionally complemented w/other data types	*** using dedicated learning algorithms (see III.2.d) with expression data optionally complemented w/other data types	** using genome-annotation, physiological information and growth phenotype data	N/A : model is typically reconstructed manually (the qualitative model may sometimes be reconstructed from data)
Identify the values of key parameters determining the behavior of the process, using experimental data	** using algorithms that learn Boolean rules compatible with prior knowledge	*** for parameters defining the probabilistic dependencies between steady-state expression levels	N/A in general (no parameters to learn); *** for metabolite concentration ranges using CBMs extended with thermodynamics constraints	*(**) depending on the size of the model to be identified. Different types of sensitivity analysis, such as Metabolic Control Analysis
Identify inconsistencies between model predictions and experiments, and use them to refine the model	** using the algorithms mentioned above (cases when identified steady state do not correspond to prior knowledge, search of missing regulatory mechanisms)	** using the algorithms mentioned above	** for regulated metabolic models using gene-expression data; ** using growth-phenotype data; ** using metabolomics data and CBMs extended with thermodynamics constraints	** using simulation and comparison with experiments: (i) with data used for model fitting, (ii) with new data
Characterize the topology of the interaction network	*** using the underlying interaction graph	*** using the underlying interaction graph	*** using the underlying interaction graph	*** using the underlying interaction graph
Predict steady-state behavior under well-defined environmental conditions	*** using steady-state/attractors analysis methods and feedback loop analysis to determine stability	N/A (or use qualitative model derived from probabilistic representation)	*** using FBA or related phenotype prediction methods	Use criteria for goodness of fit
Predict change in steady-state behavior after genetic or environmental perturbation	N/A	N/A	*** same as above, after inactivating the reactions and/or modifying exchange fluxes	** same as above, after modification of the set of ODEs (but parameters may change with genetic perturbation)
Characterize properties of the set of attainable steady-state behaviors of the process	** Use of the synchronous and asynchronous Boolean update to identify subsets of states leading to steady states	?	*** using methods such as elementary-mode analysis, flux-coupling analysis, probabilistic descriptions,	*** for small networks with known quantitative parameters, or larger networks using qualitative models
Simulate transient dynamic behavior / reaction to external stimuli	Similar as knock-out and over-expression processes described in genetic and environmental perturbation	N/A	* using dynamic FBA: growth simulations	* small networks: analytically. large networks: by applying Monte-Carlo-type sampling of parameter space; *** by simulation, in cases supported by bifurcation analysis

Table 2. Relationships between modeling frameworks and biological questions. Each column represents a model type; each row represents a type of question. The cells contain 2 types of information: 1) the degree of relevance of the modeling framework to the question (indicated by 1 to 3 stars; N/A stands for not applicable); 2) a short explanation on how the approach enables to answer the question (which method or data are used, for instance).

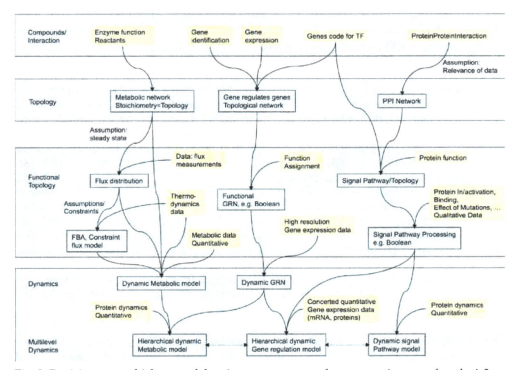

Fig. 2. Decision tree: which type of data is necessary to perform a certain type of analysis?

Given the great development computational approaches in systems biology, it has become urgent to establish methods to assess the accuracy of these in specific contexts. Challenges have been organized for example in protein structure assessment (CASP (Moult et al. 2007)) or in network reconstruction (DREAM (Stolovitzky, Monroe, and Califano 2007)). In this context, one application of models can be to challenge the topology of computationally reconstructed networks *in silico*. Models can thus both be improved with new data but also be used to test the likelihood that these data make sense in the context of the already validated model's components.

5. Examples of reconstruction procedures

5.1 T-helper model

The vertebrate immune system encompasses diverse cell populations. One of these is made of CD4+ lymphocytes, or T-helper cells. While these cells have no cytotoxic or phagocytic activity, they coordinate several cellular and humoral immune responses via the release of cytokines, which influences the activity of several cell types. In vitro, T-helper cells can be further subdivided into precursor Th0 cells and effector Th1 and Th2 cells, depending on their pattern of secreted molecules. Various mathematical models have been proposed to describe the differentiation, activation and proliferation of T-helper cells. Early models aimed to describe the cellular interactions of these cells, mediated by the secretion of cytokines. More recent models have been developed to describe the molecular mechanism that determine the differentiation process of these cells, as well as the determination of their molecular markers. There is currently a lack of quantitative data on the levels of expression of the molecules involved in the differentiation process of T-helper cells. There is, however, a vast amount of qualitative information regarding the regulatory interactions among many such molecules. As a result, it has been possible to reconstruct the basic signaling network that controls the differentiation of T-helper cells. Hence, despite the lack of quantitative data, it has been possible to model this network as a dynamical system at a qualitative level. Specifically, this network has been studied by modeling it as a discrete dynamical system (Mendoza 2006), a continuous dynamical system (Mendoza and Xenarios 2006), a Petri Net (Remy et al. 2006), and a binary decision diagram (Garg et al. 2007). Also the model has been recently updated to include the description of Th17 and Treg cell types (Mendoza and Pardo 2010). Despite the very different approaches used to model the T-helper signaling network, they all reach comparable results. Specifically, they show that the network has fixed points that correspond to the patterns of activation or expression observed in the Th0, Th1 and Th2 cell types. Furthermore, such models are capable of describing the effect of null-mutations, or over-expression of some molecules as reported by several experimental groups. The consistency among the results of several modeling approaches on the same signaling network shows that the qualitative dynamical behavior of the network is determined to a large extent by its topology. Moreover, these models show that it is possible to develop dynamical models at a qualitative level, and that such models are indeed useful to describe and predict relevant biological processes.

5.2 Genome-scale metabolic model of *Acinetobacter baylyi* ADP1

Acinetobacter baylyi ADP1, a strictly aerobic gamma-proteobacterium, is a good model organism for genetic and metabolic investigations (Metzgar et al. 2004; Young, Parke, and Ornston 2005) as well as for biotechnological applications (Abdel-El-Haleem 2003) thanks to its metabolic versatility and high competency for natural transformation. Following its sequencing and expert annotation (Barbe et al. 2004), a genome-wide collection of single-knockout mutants was generated and mutant growth phenotypes were assayed in selected environmental conditions (de Berardinis et al. 2008). In order to interpret these phenotype results and assess their consistency with the previous biochemical knowledge, a genome-scale constraint-based model of its metabolism was reconstructed (Durot et al. 2008). In a first step, an initial model was built using data from *A. baylyi*'s genome annotation, metabolic pathways databases (e.g. KEGG, BioCyc or Reactome) and physiological knowledge gathered from the literature. Such reconstruction of the overall set of biochemical reactions present in an organism has already been performed for more than 15 species (Reed et al. 2006). While facilitated by metabolism-related tools such as Pathway Tools (Karp, Paley, and Romero 2002), the task is still labor intensive and requires extensive manual curation. This initial model of *A. baylyi* included 870 reactions, 692 metabolites and 781 genes. In a second step, experimental mutant growth phenotypes were systematically compared to growth phenotypes predicted by the model. Inconsistencies between predictions and experimental results revealed potential gaps or errors in the current model: they were therefore examined to look for interpretations and possible corrections. Out of 127 inconsistencies, 60 led to a model correction. Each correction involved a modification of the gene-reaction associations, the metabolic network or the biomass requirements. Explanation of the remaining inconsistent cases would require further experimental investigations. In several cases, hypothetical interpretations involving biological processes lying outside the model scope (e.g. regulation) could be proposed. The result of that model refinement process was an increase in growth phenotype predictive accuracy from 88% to 94%. In addition, the improved version of the model integrated a significant fraction of the additional information resulting from large-scale mutant phenotyping experiments within the metabolic knowledge on *A. baylyi* derived from its genome annotation and the literature.

5.3 Nicotinamide nucleotide transhydrogenase (Nnt) pathway

Insulin secretion in mammals is carried out by pancreatic β-cells in response to the glucose concentration in the surrounding environment. The insulin secretion response is a complex one, incorporating many well-conserved biochemical reactions coupled to a cell-type-specific insulin secretion process driven by intracellular ATP concentration. The kinetic core model of pancreatic β-cell glucose-stimulated insulin secretion (Jiang, Cox, and Hancock 2007) simulates the relationship between incoming glucose concentration and resultant cytoplasmic ATP concentration using a set of ODEs representing 44 enzymatic reactions, some of which take place in more than one compartment and simulates the concentrations of 59 metabolites. The model is divided into three compartments: cytoplasm, mitochondrial matrix and mitochondrial inter-membrane space. Information on components of the model was collected from the literature and from SABIO-RK (http://www.sabio.villa-bosch.de/SABIORK/). Initial validation of the model was based on its ability to replicate published properties of the system in vivo, such as response to changing glucose concentration and oscillation of the concentration of certain metabolites in the glycolysis pathway and elsewhere. Further characterization and optimization will require *in vivo* measurement of some of these concentrations in an appropriate cell type.

6. Discussion

We conclude along the lines of the famous quote of G.E.P. Box: "All Models Are Wrong But Some Models Are Useful" (Segal et al. 2003). Most models are often focused on particular scientific domains and integrate a limited set of information, considered being "good-enough" to answer the scientific questions of a given project. Many uncertainties still remain in the interpretation of computational models. Those uncertainties emanate from different levels of variation: technology, laboratory-to-laboratory, human-to-human specific, among others. The true biological variation or stochasticity is harder to detect. For example, when there is significant variation between individual runs of an experiment, how should the data be integrated in the model? Should individual models be created for each case, and these compared, or should data be averaged to produce a consensus model? When is a model considered to contain most of the behaviors of the real system?

The different examples represented in this review are not exhaustive by far and only reflect our limited knowledge of the interface between mechanistic modeling and "wet" experiments, based on the research performed within the ENFIN Consortium.

Ultimately providing both models and experimental data to a wider community is a way to bridge the gap and transmit our knowledge to the next generation of scientists. An effort leading in that direction is the development of databases of models such as BioModels (www.biomodels.org) (Le Novere et al. 2006), which collects molecular models by using a standardized format. Another important development is the formulation of standards for experiments and for model formulation (such as MIRIAM (Le Novere et al. 2005)). Standards and associated controlled vocabularies will certainly contribute to unifying experimental data models that are currently scattered across several databases.

7. Acknowledgements

This work was supported by ENFIN, a Network of Excellence funded by the European Commission within its FP6 Programme, under the thematic area "Life sciences, genomics and biotechnology for health", contract number LSHG-CT-2005-518254.

8. References

Abdel-El-Haleem, D. 2003. Acinetobacter: environmental and biotechnological applications. *Afr. J. Biotechnol.* 2 (4):71-4.

Albert, R., and H. G. Othmer. 2003. The topology of the regulatory interactions predicts the expression pattern of the segment polarity genes in Drosophila melanogaster. *J Theor Biol* 223 (1):1-18.

Ashyraliyev, M., J. Jaeger, and J. G. Blom. 2008. Parameter estimation and determinability analysis applied to Drosophila gap gene circuits. *BMC Syst Biol* 2:83.

Barbe, V., D. Vallenet, N. Fonknechten, A. Kreimeyer, S. Oztas, L. Labarre, S. Cruveiller, C. Robert, S. Duprat, P. Wincker, L. N. Ornston, J. Weissenbach, P. Marliere, G. N. Cohen, and C. Medigue. 2004. Unique features revealed by the genome sequence of Acinetobacter sp. ADP1, a versatile and naturally transformation competent bacterium. *Nucleic Acids Res* 32 (19):5766-79.

Barkai, N., and S. Leibler. 1997. Robustness in simple biochemical networks. *Nature* 387 (6636):913-7.

Briggs, G. E., and J. B. Haldane. 1925. A Note on the Kinetics of Enzyme Action. *Biochem J* 19 (2):338-9.

Cannon, W. B. 1929. Organization for physiological homeostasis. *Physiol. Rev.* 9:399-431.

Chang, A., M. Scheer, A. Grote, I. Schomburg, and D. Schomburg. 2009. BRENDA, AMENDA and FRENDA the enzyme information system: new content and tools in 2009. *Nucleic Acids Res* 37 (Database issue):D588-92.

Chen, K. C., L. Calzone, A. Csikasz-Nagy, F. R. Cross, B. Novak, and J. J. Tyson. 2004. Integrative analysis of cell cycle control in budding yeast. *Mol Biol Cell* 15 (8):3841-62.

Cornish-Bowden, A., M. L. Cardenas, J. C. Letelier, J. Soto-Andrade, and F. G. Abarzua. 2004. Understanding the parts in terms of the whole. *Biol Cell* 96 (9):713-7.

Davidich, M. I., and S. Bornholdt. 2008. Boolean network model predicts cell cycle sequence of fission yeast. *PLoS ONE* 3 (2):e1672.

de Berardinis, V., D. Vallenet, V. Castelli, M. Besnard, A. Pinet, C. Cruaud, S. Samair, C. Lechaplais, G. Gyapay, C. Richez, M. Durot, A. Kreimeyer, F. Le Fevre, V. Schachter, V. Pezo, V. Doring, C. Scarpelli, C. Medigue, G. N. Cohen, P. Marliere, M. Salanoubat, and J. Weissenbach. 2008. A complete collection of single-gene deletion mutants of Acinetobacter baylyi ADP1. *Mol Syst Biol* 4:174.

De Jong, H., J. Geiselmann, G. Batt, C. Hernandez, and M. Page. 2004. Qualitative simulation of the initiation of sporulation in Bacillus subtilis. *Bull Math Biol* 66 (2):261-99.

Delbrück, M. 1949. *Discussion In: Unités Biologiques Douées de Continuité Génétique* Lyon: CNRS.
Durot, M., P. Y. Bourguignon, and V. Schachter. 2009. Genome-scale models of bacterial metabolism: reconstruction and applications. *FEMS microbiology reviews* 33 (1):164-90.
Durot, M., F. Le Fevre, V. de Berardinis, A. Kreimeyer, D. Vallenet, C. Combe, S. Smidtas, M. Salanoubat, J. Weissenbach, and V. Schachter. 2008. Iterative reconstruction of a global metabolic model of Acinetobacter baylyi ADP1 using high-throughput growth phenotype and gene essentiality data. *BMC systems biology* 2:85.
Ellis, R. J. 2001. Macromolecular crowding: obvious but underappreciated. *Trends Biochem Sci* 26 (10):597-604.
Erdi, P., and J. Toth. 1989. *Mathematical Models of Chemical Reactions*. Manchester: Manchester University Press.
Erfle, H., B. Neumann, U. Liebel, P. Rogers, M. Held, T. Walter, J. Ellenberg, and R. Pepperkok. 2007. Reverse transfection on cell arrays for high content screening microscopy. *Nat Protoc* 2 (2):392-9.
Faure, A., A. Naldi, C. Chaouiya, and D. Thieffry. 2006. Dynamical analysis of a generic Boolean model for the control of the mammalian cell cycle. *Bioinformatics* 22 (14):e124-31.
Feist, A. M., M. J. Herrgard, I. Thiele, J. L. Reed, and B. O. Palsson. 2009. Reconstruction of biochemical networks in microorganisms. *Nature reviews. Microbiology* 7 (2):129-43.
Feist, A. M., and B. O. Palsson. 2008. The growing scope of applications of genome-scale metabolic reconstructions using Escherichia coli. *Nature Biotechnology* 26 (6):659-67.
Ferrell, J. E., Jr. 2002. Self-perpetuating states in signal transduction: positive feedback, double-negative feedback and bistability. *Curr Opin Cell Biol* 14 (2):140-8.
Fournier, T., J. P. Gabriel, C. Mazza, J. Pasquier, J. L. Galbete, and N. Mermod. 2007. Steady-state expression of self-regulated genes. *Bioinformatics* 23 (23):3185-92.
Friedman, N., M. Linial, I. Nachman, and D. Pe'er. 2000. Using Bayesian networks to analyze expression data. *J Comput Biol* 7 (3-4):601-20.
Garg, A., D. Banerjee, and G. De Micheli. 2008. Implicit methods for probabilistic modeling of Gene Regulatory Networks. *Conf Proc IEEE Eng Med Biol Soc* 2008:4621-7.
Garg, A., I. Xenarios, L. Mendoza, and G. DeMicheli. 2007. An efficient method for dynamic analysis of gene regulatory networks and in silico gene perturbation experiments. *RECOMB 2007, Lecture Notes in Computer Science* 4453:62-76.
Gasch, A. P., P. T. Spellman, C. M. Kao, O. Carmel-Harel, M. B. Eisen, G. Storz, D. Botstein, and P. O. Brown. 2000. Genomic expression programs in the response of yeast cells to environmental changes. *Mol Biol Cell* 11 (12):4241-57.
Glass, L., and S. A. Kauffman. 1973. The logical analysis of continuous, non-linear biochemical control networks. *J Theor Biol* 39 (1):103-29.
Glass, L., and JS Pasternack. 1978. Stable oscillations in mathematical models of biological control systems. *J Math Biol* 6:207-223.
Goldbeter, A. 1991. A minimal cascade model for the mitotic oscillator involving cyclin and cdc2 kinase. *Proc Natl Acad Sci U S A* 88 (20):9107-11.
Gonzalez, A. G., A. Naldi, L. Sanchez, D. Thieffry, and C. Chaouiya. 2006. GINsim: a software suite for the qualitative modelling, simulation and analysis of regulatory networks. *Biosystems* 84 (2):91-100.

Gouzé, J. L. 1998. Positive and negative circuits in dynamical systems. *J. Biol. Systems* 6:11-15.

Gouzé, J. L., and T. Sari. 2002. A class of piecewise linear differential equations arising in biological models. *Dyn. Syst.* 17:299-316.

Guldberg, C. M., and P. Waage. 1879. Concerning Chemical Affinity. *Erdmann's Journal für Practische Chemie* 127:69-114.

Gupta, S., S. S. Bisht, R. Kukreti, S. Jain, and S. K. Brahmachari. 2007. Boolean network analysis of a neurotransmitter signaling pathway. *J Theor Biol* 244 (3):463-9.

Heidtke, K. R., and S. Schulze-Kremer. 1998. Design and implementation of a qualitative simulation model of lambda phage infection. *Bioinformatics* 14 (1):81-91.

Heinrich, R., and T. A. Rapoport. 1974. A linear steady-state treatment of enzymatic chains. General properties, control and effector strength. *Eur J Biochem* 42 (1):89-95.

Henry, C. S., M. DeJongh, A. A. Best, P. M. Frybarger, B. Linsay, and R. L. Stevens. 2010. High-throughput generation, optimization and analysis of genome-scale metabolic models. *Nature Biotechnology* 28 (9):977-82.

Henry, C. S., M. D. Jankowski, L. J. Broadbelt, and V. Hatzimanikatis. 2006. Genome-scale thermodynamic analysis of Escherichia coli metabolism. *Biophys J* 90 (4):1453-61.

Higgins, J. 1964. A Chemical Mechanism for Oscillation of Glycolytic Intermediates in Yeast Cells. *Proc Natl Acad Sci U S A* 51:989-94.

Huang, S., and D. E. Ingber. 2000. Shape-dependent control of cell growth, differentiation, and apoptosis: switching between attractors in cell regulatory networks. *Exp Cell Res* 261 (1):91-103.

Hubbard, T. J., B. L. Aken, K. Beal, B. Ballester, M. Caccamo, Y. Chen, L. Clarke, G. Coates, F. Cunningham, T. Cutts, T. Down, S. C. Dyer, S. Fitzgerald, J. Fernandez-Banet, S. Graf, S. Haider, M. Hammond, J. Herrero, R. Holland, K. Howe, K. Howe, N. Johnson, A. Kahari, D. Keefe, F. Kokocinski, E. Kulesha, D. Lawson, I. Longden, C. Melsopp, K. Megy, P. Meidl, B. Ouverdin, A. Parker, A. Prlic, S. Rice, D. Rios, M. Schuster, I. Sealy, J. Severin, G. Slater, D. Smedley, G. Spudich, S. Trevanion, A. Vilella, J. Vogel, S. White, M. Wood, T. Cox, V. Curwen, R. Durbin, X. M. Fernandez-Suarez, P. Flicek, A. Kasprzyk, G. Proctor, S. Searle, J. Smith, A. Ureta-Vidal, and E. Birney. 2007. Ensembl 2007. *Nucleic Acids Res* 35 (Database issue):D610-7.

Hynne, F., S. Dano, and P. G. Sorensen. 2001. Full-scale model of glycolysis in Saccharomyces cerevisiae. *Biophys Chem* 94 (1-2):121-63.

Jaeger, J., S. Surkova, M. Blagov, H. Janssens, D. Kosman, K. N. Kozlov, Manu, E. Myasnikova, C. E. Vanario-Alonso, M. Samsonova, D. H. Sharp, and J. Reinitz. 2004. Dynamic control of positional information in the early Drosophila embryo. *Nature* 430 (6997):368-71.

Jiang, N., R. D. Cox, and J. M. Hancock. 2007. A kinetic core model of the glucose-stimulated insulin secretion network of pancreatic beta cells. *Mamm Genome* 18 (6-7):508-520.

Joyce, A. R., and B. O. Palsson. 2006. The model organism as a system: integrating 'omics' data sets. *Nat Rev Mol Cell Biol* 7 (3):198-210.

Kacser, H., and J. A. Burns. 1973. The control of flux. *Symp Soc Exp Biol* 27:65-104.

Kanehisa, M., S. Goto, M. Hattori, K. F. Aoki-Kinoshita, M. Itoh, S. Kawashima, T. Katayama, M. Araki, and M. Hirakawa. 2006. From genomics to chemical genomics: new developments in KEGG. *Nucleic Acids Res* 34 (Database issue):D354-7.

Karp, P. D., S. Paley, and P. Romero. 2002. The Pathway Tools software. *Bioinformatics* 18 Suppl 1:S225-32.
Kauffman, S. A. 1969. Metabolic stability and epigenesis in randomly constructed genetic nets. *J Theor Biol* 22 (3):437-67.
Repeated Author. 1993. *The origins of order: Self-organization and selection in evolution*: Oxford University Press.
Kerrien, S., Y. Alam-Faruque, B. Aranda, I. Bancarz, A. Bridge, C. Derow, E. Dimmer, M. Feuermann, A. Friedrichsen, R. Huntley, C. Kohler, J. Khadake, C. Leroy, A. Liban, C. Lieftink, L. Montecchi-Palazzi, S. Orchard, J. Risse, K. Robbe, B. Roechert, D. Thorneycroft, Y. Zhang, R. Apweiler, and H. Hermjakob. 2007. IntAct--open source resource for molecular interaction data. *Nucleic Acids Res* 35 (Database issue):D561-5.
Kervizic, G., and L. Corcos. 2008. Dynamical modeling of the cholesterol regulatory pathway with Boolean networks. *BMC Syst Biol* 2 (1):99.
Klipp, E. 2007. Modelling dynamic processes in yeast. *Yeast*.
Klipp, E., R. Herwig, A. Kowald, C. Wierling, and H. Lehrach. 2005. *Systems Biology in Practice*. Edited by Wiley-VCH.
Koshland, D. E., Jr., G. Nemethy, and D. Filmer. 1966. Comparison of experimental binding data and theoretical models in proteins containing subunits. *Biochemistry* 5 (1):365-85.
Kummel, A., S. Panke, and M. Heinemann. 2006. Putative regulatory sites unraveled by network-embedded thermodynamic analysis of metabolome data. *Mol Syst Biol* 2:2006 0034.
Le Novere, N., B. Bornstein, A. Broicher, M. Courtot, M. Donizelli, H. Dharuri, L. Li, H. Sauro, M. Schilstra, B. Shapiro, J. L. Snoep, and M. Hucka. 2006. BioModels Database: a free, centralized database of curated, published, quantitative kinetic models of biochemical and cellular systems. *Nucleic Acids Res* 34 (Database issue):D689-91.
Le Novere, N., A. Finney, M. Hucka, U. S. Bhalla, F. Campagne, J. Collado-Vides, E. J. Crampin, M. Halstead, E. Klipp, P. Mendes, P. Nielsen, H. Sauro, B. Shapiro, J. L. Snoep, H. D. Spence, and B. L. Wanner. 2005. Minimum information requested in the annotation of biochemical models (MIRIAM). *Nat Biotechnol* 23 (12):1509-15.
Lewis, JE, and L. Glass. 1995. Steady states, limit cycles, and chaos in models of complex biological networks. *Int J Bifurcat Chaos* 1:477-483.
Li, F., T. Long, Y. Lu, Q. Ouyang, and C. Tang. 2004. The yeast cell-cycle network is robustly designed. *Proc Natl Acad Sci U S A* 101 (14):4781-6.
Liebermeister, W., and E. Klipp. 2006. Bringing metabolic networks to life: convenience rate law and thermodynamic constraints. *Theor Biol Med Model* 3:41.
Matthews, L., G. Gopinath, M. Gillespie, M. Caudy, D. Croft, B. de Bono, P. Garapati, J. Hemish, H. Hermjakob, B. Jassal, A. Kanapin, S. Lewis, S. Mahajan, B. May, E. Schmidt, I. Vastrik, G. Wu, E. Birney, L. Stein, and P. D'Eustachio. 2009. Reactome knowledgebase of human biological pathways and processes. *Nucleic Acids Res* 37 (Database issue):D619-22.
Mendoza, L. 2006. A network model for the control of the differentiation process in Th cells. *Biosystems* 84 (2):101-14.
Mendoza, L., and F. Pardo. 2010. A robust model to describe the differentiation of T-helper cells. *Theory in biosciences = Theorie in den Biowissenschaften* 129 (4):283-93.

Mendoza, L., D. Thieffry, and E. R. Alvarez-Buylla. 1999. Genetic control of flower morphogenesis in Arabidopsis thaliana: a logical analysis. *Bioinformatics* 15 (7-8):593-606.

Mendoza, L., and I. Xenarios. 2006. A method for the generation of standardized qualitative dynamical systems of regulatory networks. *Theor Biol Med Model* 3:13.

Mestl, T., RJ Bagley, and L. Glass. 1997. Common chaos in arbitrarily complex feedback networks. *Phys. Rev. Letters* 79:653-656.

Mestl, T., E. Plahte, and S. W. Omholt. 1995. A mathematical framework for describing and analysing gene regulatory networks. *J Theor Biol* 176 (2):291-300.

Metzgar, D., J. M. Bacher, V. Pezo, J. Reader, V. Doring, P. Schimmel, P. Marliere, and V. de Crecy-Lagard. 2004. Acinetobacter sp. ADP1: an ideal model organism for genetic analysis and genome engineering. *Nucleic Acids Res* 32 (19):5780-90.

Michaelis, L., and M. Menten. 1913. Die Kinetik der Invertinwirkung. *Biochem. Z.* 49:333-369.

Monod, J., J. Wyman, and J. P. Changeux. 1965. On the Nature of Allosteric Transitions: a Plausible Model. *J Mol Biol* 12:88-118.

Morgan, H., T. Beck, A. Blake, H. Gates, N. Adams, G. Debouzy, S. Leblanc, C. Lengger, H. Maier, D. Melvin, H. Meziane, D. Richardson, S. Wells, J. White, J. Wood, M. H. de Angelis, S. D. Brown, J. M. Hancock, and A. M. Mallon. 2010. EuroPhenome: a repository for high-throughput mouse phenotyping data. *Nucleic Acids Res* 38 (Database issue):D577-85.

Moult, J., K. Fidelis, A. Kryshtafovych, B. Rost, T. Hubbard, and A. Tramontano. 2007. Critical assessment of methods of protein structure prediction-Round VII. *Proteins* 69 Suppl 8:3-9.

Papin, J. A., J. Stelling, N. D. Price, S. Klamt, S. Schuster, and B. O. Palsson. 2004. Comparison of network-based pathway analysis methods. *Trends Biotechnol* 22 (8):400-5.

Parkinson, H., M. Kapushesky, M. Shojatalab, N. Abeygunawardena, R. Coulson, A. Farne, E. Holloway, N. Kolesnykov, P. Lilja, M. Lukk, R. Mani, T. Rayner, A. Sharma, E. William, U. Sarkans, and A. Brazma. 2007. ArrayExpress--a public database of microarray experiments and gene expression profiles. *Nucleic Acids Res* 35 (Database issue):D747-50.

Pearl, J. 1988. *Probabilistic Reasoning in Intelligent Systems: Networks of Plausible Inference*. Edited by M. Kaufmann. San Mateo, CA.

Plahte, E., T. Mestl, and S.W. Omholt. 1994. Global analysis of steady points for systems of differential equations with sigmoid interactions. *Dyn Stabil Syst* 9:275-291.

Plahte, E., T. Mestl, and S.W. Omholt. 1995. Feedback loops, stability and multistationarity in dynamical systems. *J. Biol. Systems* 3:409-413.

Reed, J. L., I. Famili, I. Thiele, and B. O. Palsson. 2006. Towards multidimensional genome annotation. *Nat Rev Genet* 7 (2):130-41.

Remy, E., P. Ruet, L. Mendoza, D. Thieffry, and C. Chaouiya. 2006. From logical regulatory graphs to standard petri nets: Dynamical roles and functionality of feedback circuits. *Transactions on Computational Systems Biology VII, Lecture Notes in Computer Science* 4230:56-72.

Rojas, I., M. Golebiewski, R. Kania, O. Krebs, S. Mir, A. Weidemann, and U. Wittig. 2007. Storing and annotating of kinetic data. *In Silico Biol* 7 (2 Suppl):S37-44.

Ropers, D., H. de Jong, M. Page, D. Schneider, and J. Geiselmann. 2006. Qualitative simulation of the carbon starvation response in Escherichia coli. *Biosystems* 84 (2):124-52.

Saez-Rodriguez, J., L. Simeoni, J. A. Lindquist, R. Hemenway, U. Bommhardt, B. Arndt, U. U. Haus, R. Weismantel, E. D. Gilles, S. Klamt, and B. Schraven. 2007. A logical model provides insights into T cell receptor signaling. *PLoS Comput Biol* 3 (8):e163.

Samal, A., and S. Jain. 2008. The regulatory network of E. coli metabolism as a Boolean dynamical system exhibits both homeostasis and flexibility of response. *BMC Syst Biol* 2:21.

Sanchez-Corrales, Y. E., E. R. Alvarez-Buylla, and L. Mendoza. 2010. The Arabidopsis thaliana flower organ specification gene regulatory network determines a robust differentiation process. *Journal of theoretical biology* 264 (3):971-83.

Schellenberger, J., and B. O. Palsson. 2009. Use of randomized sampling for analysis of metabolic networks. *The Journal of biological chemistry* 284 (9):5457-61.

Segal, E., M. Shapira, A. Regev, D. Pe'er, D. Botstein, D. Koller, and N. Friedman. 2003. Module networks: identifying regulatory modules and their condition-specific regulators from gene expression data. *Nat Genet* 34 (2):166-76.

Segal, E., R. Yelensky, and D. Koller. 2003. Genome-wide discovery of transcriptional modules from DNA sequence and gene expression. *Bioinformatics* 19 Suppl 1:i273-82.

Sel'kov, E. E. 1968. Self-oscillations in glycolysis. 1. A simple kinetic model. *Eur J Biochem* 4 (1):79-86.

Shlomi, T., M. N. Cabili, M. J. Herrgard, B. O. Palsson, and E. Ruppin. 2008. Network-based prediction of human tissue-specific metabolism. *Nature Biotechnology* 26 (9):1003-10.

Snoussi, E. H. 1998. Necessary conditions for multistationarity and stable periodicity. *J. Biol. Systems* 6:3-9.

Sorribas, A., and M. Cascante. 1994. Structure identifiability in metabolic pathways: parameter estimation in models based on the power-law formalism. *Biochem J* 298 (Pt 2):303-11.

Stolovitzky, G., D. Monroe, and A. Califano. 2007. Dialogue on reverse-engineering assessment and methods: the DREAM of high-throughput pathway inference. *Ann N Y Acad Sci* 1115:1-22.

Tavazoie, S., J. D. Hughes, M. J. Campbell, R. J. Cho, and G. M. Church. 1999. Systematic determination of genetic network architecture. *Nat Genet* 22 (3):281-5.

Teusink, B., J. Passarge, C. A. Reijenga, E. Esgalhado, C. C. van der Weijden, M. Schepper, M. C. Walsh, B. M. Bakker, K. van Dam, H. V. Westerhoff, and J. L. Snoep. 2000. Can yeast glycolysis be understood in terms of in vitro kinetics of the constituent enzymes? Testing biochemistry. *Eur J Biochem* 267 (17):5313-29.

The UniProt Consortium. 2007. The Universal Protein Resource (UniProt). *Nucleic Acids Res* 35 (Database issue):D193-7.

Thieffry, D., E. H. Snoussi, J. Richelle, and R. Thomas. 1995. Positive loops and differentiation. *J. Biol. Systems* 3:457-466.

Thomas, R. 1973. Boolean formalization of genetic control circuits. *J Theor Biol* 42 (3):563-85.

Trelease, R. B., R. A. Henderson, and J. B. Park. 1999. A qualitative process system for modeling NF-kappaB and AP-1 gene regulation in immune cell biology research. *Artif Intell Med* 17 (3):303-21.

Tyson, J. J., K. C. Chen, and B. Novak. 2003. Sniffers, buzzers, toggles and blinkers: dynamics of regulatory and signaling pathways in the cell. *Curr Opin Cell Biol* 15 (2):221-31.

Tyson, J. J., A. Csikasz-Nagy, and B. Novak. 2002. The dynamics of cell cycle regulation. *Bioessays* 24 (12):1095-109.

Varma, A., and B. O. Palsson. 1994. Stoichiometric flux balance models quantitatively predict growth and metabolic by-product secretion in wild-type Escherichia coli W3110. *Appl Environ Microbiol* 60 (10):3724-31.

Viretta, A. U., and M. Fussenegger. 2004. Modeling the quorum sensing regulatory network of human-pathogenic Pseudomonas aeruginosa. *Biotechnol Prog* 20 (3):670-8.

Waage, P., and C. M. Guldberg. 1864. Studies Concerning Affinity. *Forhandlinger: Videnskabs – Selskabet i Christinia* 35.

Wolf, J., and R. Heinrich. 2000. Effect of cellular interaction on glycolytic oscillations in yeast: a theoretical investigation. *Biochem J* 345 Pt 2:321-34.

Wolf, J., J. Passarge, O. J. Somsen, J. L. Snoep, R. Heinrich, and H. V. Westerhoff. 2000. Transduction of intracellular and intercellular dynamics in yeast glycolytic oscillations. *Biophys J* 78 (3):1145-53.

Young, D. M., D. Parke, and L. N. Ornston. 2005. Opportunities for genetic investigation afforded by Acinetobacter baylyi, a nutritionally versatile bacterial species that is highly competent for natural transformation. *Annu Rev Microbiol* 59:519-51.

Zi, Z., and E. Klipp. 2007. Constraint-based modeling and kinetic analysis of the smad dependent tgf-Beta signaling pathway. *PLoS ONE* 2 (9):e936.

Zimmerman, S. B., and A. P. Minton. 1993. Macromolecular crowding: biochemical, biophysical, and physiological consequences. *Annu Rev Biophys Biomol Struct* 22:27-65.

Signal Processing Methods for Capillary Electrophoresis

Robert Stewart[1], Iftah Gideoni[2] and Yonggang Zhu[1]
[1]*CSIRO Materials Science and Engineering,*
[2]*CSIRO Information and Communication Technology Centre,*
Australia

1. Introduction

Capillary electrophoresis (CE) is a separation technique that can be used as a sample pre-treatment step in the analysis of ionic analytes (Grossman and Colburn 1992; Stewart et al. 2008). Compared with other separation technologies it can offer advantages such as higher speed and sensitivity, smaller injection volumes and reduced consumption of solvent and samples, the possibility of miniaturisation, and reduced cost (Issaq 2001; Jarméus and Emmer 2008; Polesello and Valsecchi 1999; Wang 2005; Wee et al. 2008).

CE is based on the difference of the electrical mobilities of molecules within a capillary tube filled with electrolyte solution. When an electrical field is applied between the two ends of a capillary and a sample is introduced at one end, analytes are separated as they migrate towards the other end under the influence of the electrical field. These separated analytes are detected near the outlet by methods such as optical or electrochemical techniques (Polesello and Valsecchi 1999; Guijt et al. 2004; Kappes and Hauser 1999; Kubáň and Hauser 2004, 2009; Kuhn and Hoffstetter-Kuhn 1993; Marzilli et al. 1997; Tanyanyiwa et al. 2002; Zemann et al. 1998). The signal from a detector is digitised and typically presented in the form of voltage versus time, i.e. an *electropherogram*. Peaks evident in an electropherogram typically correspond to analytes in the sample, and with optimisation of the system parameters, the peaks can usually be resolved sufficiently. Fig. 1 shows an example electropherogram of data obtained from a practical trial reported earlier (Petkovic-Duran et al. 2008)

For analytical chemistry purposes, the operator's aim is to determine from the electropherogram what analytes are present and the corresponding concentrations. In this paper we assume that this is done by separating the task into two stages: Signal Processing, i.e., obtaining peak information from the electropherogram, and Pattern Matching, using this peaks' summary information to compare with established peak library of known chemicals. Whilst the process of identifying the peaks, removing the noise present and fitting curves for peak quantification is, to a large extent, done to-date manually by professionals, the operator would be greatly aided through fully automated techniques with little or no human input. This is particularly crucial for the development of field-deployable devices which could be operated by non-technical staff. Furthermore, automated signal processing techniques can allow results to be reproducible or consistent and can remove the subjectivity of a human evaluation. In addition, they can also detect features that may not otherwise be obvious to the human eye, and enable the operation of detection devices by non-experts.

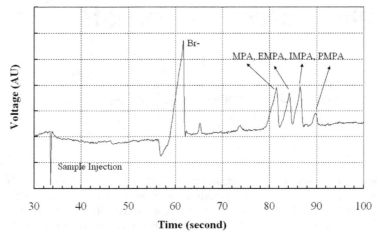

Fig. 1. An example electropherogram for separating a mixture of chemical warfare agent degradation products MPA, EMPA, IMPA and PMPA. Reproduced from Petkovic-Duran (2008). Conditions: 10mM MES/His (pH 6.1) buffer; separation field strength 340V/cm; injection field strength 625V/cm; Injection time 20s; frequency 360kH; peak-to-peak-voltage 8V; sinus ac waveform. The first spike corresponds to the start of sample injection into the CE channel.

The task of automating the extraction of peaks, i.e. obtaining peak information such as peak shape, peak height, peak area and arrival time, from an electropherogram is, however, not an easy one. Not only must signal processing techniques be developed that can find the location of peaks, but they must do so in the presence of low and high frequency baseline noise that corrupts the signal. Furthermore, analytes may co-elute resulting in poorly resolved peaks that overlap. Even once the peak locations have been identified, the peaks need to be extracted and/or accurately quantified in the presence of interfering signal components.

The purpose of this paper is to review the current progress in signal processing relevant for capillary electrophoresis that are directed towards the quantification of peaks in electropherograms. We provide an overview of a signal processing strategy for a complete system, and then detail each signal processing step through examples cited from the literature. We are then able to draw some conclusions about the work needed to develop well defined and completely automated signal processing systems for CE. In the next section (Section 3) we detail a model for the electropherogram signal and the signal components to be analysed. Section 4 provides an overview of the steps in the signal processing strategy for CE signals. Pertinent examples from the literature are cited for each step (baseline noise removal, peak finding and peak extraction and quantification) to illustrate the different approaches adopted. Section 5 addresses how the performance of signal processing strategies/algorithms should be assessed, and on the need for benchmark testing and for the provision of system specifications. Section 6 discuss briefly some difficulties and requirements for the future Pattern Matching work needed for practically extracting chemical identity information from the electropherogram's peak data. A concluding summary is provided in Section 7.

2. Modelling an electropherogram signal

An electropherogram is typically modelled as the superposition of a number of components under the assumption of system linearity (as are chromatograms (Dyson 1998) and mass spectrograms (Coombes et al. 2005)). These include the peaks corresponding to the analytes, system peaks and noise components. Mathematically, the model for an unfiltered/raw electropherogram signal can be expressed, e.g., in the form of voltage $v(t)$ (Kubáň and Hauser 2009), as follows,

$$v(t) = \sum_{i=1}^{N} p_i(t) + \sum_{j=1}^{M} s_j(t) + B + n(t) \qquad (1)$$

where $p_i(t)$ is a peak that corresponds to the ith analyte eluted, N is the number of analyte peaks, $s_j(t)$ is the jth system peak, M is the number of system peaks, B is a constant baseline, and $n(t)$ is unwanted baseline noise. Here we consider a constant baseline, and any deviation from this is due to baseline noise, $n(t)$, which may itself contain a number of components. Eq. (1) is useful, not only for modelling electropherograms obtained from physical systems, but also for devising synthetic signals to test peak finding or peak extraction algorithms.

The aim of a peak extraction algorithm is to apply signal processing means to extract the separate peak components, $p_i(t)$, from an electropherogram. Each peak component can then be quantitatively analysed to provide information pertaining to the concentration of its corresponding analyte. When the sample being tested is unknown, the information obtained for the peaks can then be used in conjunction with prior knowledge or a database to identify the chemical compounds present in the sample (Reichenbach et al. 2009). In order to complete these tasks successfully, it is important to model or understand the characteristics of all the signal components in Eq. (1) and this will be discussed in detail in the following sections.

2.1 Peak models

Before discussing peak models, it is important to clarify the terminologies associated with a peak. On the basis of the peak definitions of The International Union of Pure and Applied Chemistry (IUPAC) and the Differential Thermal Analysis (DTA) (Inczédy et al. 1998), we propose a general definition for a peak with a view to it being widely applicable in different situations within a CE context, i.e. a peak is a portion of a signal or waveform with the characteristic of a rising and then a falling of the dependent variable with time. In particular, an analyte peak is a peak that is a signal component of an electropherogram (viz. $p_i(t)$ in Eq. (1)) resulting directly from the presence of an analyte and distinct from noise and other peaks [e.g. see (Vivó-Truyols et al. 2005a)] while a system peak is a peak directly resulting from the background electrolyte (viz. $s_j(t)$ in Eq. (1)) (Beckers and Everaerts 1997; Gaš et al. 2007; Gebauer and Boček 1997; Macka et al. 1997; Sellmeyer and Poppe 2002).

Fig. 2 provides an illustrative example of the qualitative definitions above. It can be seen how system and analyte peaks contribute to the electropherogram signal which also contains low and high frequency baseline noise. It should be noted that we are only considering peaks that are at a coarser scale than the high frequency noise that is present. In this figure it is clear that if the low and high frequency baseline noise were removed, then the electropherogram peaks would approximate the analytes or system peaks provided they are fully resolved (don't overlap) and have the same constant baseline. It should also be mentioned that a peak can be either 'positive' or 'negative'. Negatives peaks correspond to changes to below baseline and appear as valleys in an electropherogram. The profile of an analyte peak is dependent ``on the physical-chemical processes inside the capillary, the heterogeneity of the capillary surface, capillary overload, solute mobility, and instrumental effects''(García-Alvarez-Coque et al. 2005).

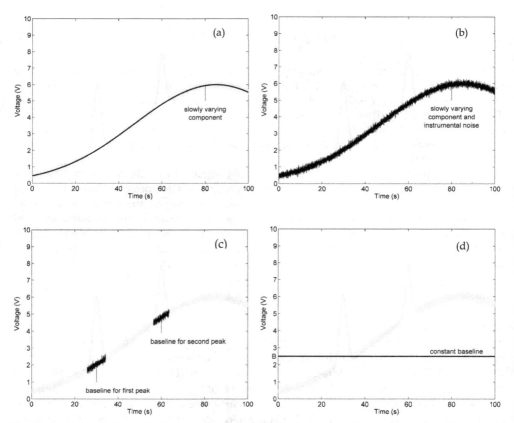

Fig. 2. A synthetic electropherogram composed from the superposition of peaks and baseline noise. Each interval in the electropherogram shown in bold corresponds to a successful or unsuccessful peak candidate.

A number of different peak models have been proposed in the literature. The triangle is likely to be the earliest peak model used (Dyson 1998) and is perhaps the simplest. It has been used as a peak model in a number of studies (Barclay and Bonner 1997; Stewart et al. 2008). One definition for a triangle function is (Couch 1990),

$$\Lambda\left(\frac{t}{T}\right) = \begin{cases} 1 - \frac{|t|}{T}, & |t| \leq T \\ 0, & |t| > T \end{cases} \quad (2)$$

where t is time, and T is half the width of the triangle which has unity height and is centred about the y-axis with an apex at (0,1). This function can be scaled, translated and sampled to give a suitable digital signal representation of a peak.

A more common approach, however, has been to model a peak as a Gaussian curve or variant thereof (Bocaz-Beneventi et al. 2002; Oda and Landers 19970. Such curves are likely to be more realistic given the underlying physical-chemical process. For example, Solis et al. (Solis et al. 2007) define a peak to be of the form (Graves-Morris et al. 2006):

$$p_i[t] = A_i \exp\left[-4\left(\frac{T_S t - T o_i}{W_i}\right)\right] \quad (3)$$

where A_i is the peak's amplitude, To_i is the migration time, W_i the width of the peak and T_S the sampling interval. This model could be applied to both $p_i(t)$ and $s_j(t)$ in Eq. (1). More complex peak models have also been introduced for CE. For example, to account for skewed peaks, a ``Combination of Square Roots (CSR)" model has been proposed (García-Alvarez-Coque et al. 2005) and compared to other models. Another example is the resonance model for peaks described in (Graves-Morris et al. 2006).

As with CE, there is a need for peak models in other analytical chemistry techniques including chromatography, spectroscopy and gel and zone electrophoresis. The signals obtained in these techniques are generally alike, often owing to the similarity between the underlying physical-chemical processes (the mechanisms in chromatography and electrophoresis are particularly similar (Johansson et al. 2003; Poppe 1999). Gaussian based peak models are a popular choice for various techniques in analytical chemistry. For example, in chromatography the peaks in chromatograms are expected in theory to be close to Gaussian as they result from the dispersion of sample bands (Dyson 1998; Parris 1984). However, in practice, as with CE, Gaussian models must be modified or replaced to capture other effects that impact on peak shape (including asymmetry). Examples of modified Gaussian peak models include Exponentially Modified Gaussian (EMG) (Grushka 1972; Naish and Hartwell 1988; Poole 2003) and Exponential-Gaussian Hybrid (EGH) functions (Lan and Jorgenson 2001; Poole 2003). Numerous other peak models for chromatography have also been proposed (see the references cited in (García-Alvarez-Coque et al. 2005) for some further examples).

2.2 Baseline noise

Noise is usually the result of "spontaneous fluctuations which occur within matter at the microscopic level" (e.g. thermal noise and shot noise). In a CE system, noise components in a baseline signal may be due to electrical noise, chemical noise originating in the underlying physical-chemical processes of separation and so on. In general, baseline noise could contain both low frequency and high frequency components. The high frequency noise is from the instrument/detector which results from ``incomplete grounding or from the signal amplification system" (Kuhn and Hoffstetter-Kuhn 1993) or from other sources such as electronic components including the Analogue to Digital Converter (ADC) (Jacobsen 2007;

Solis et al. 2007; Xu et al. 1997). The low frequency noise is mainly generated from temperature variations, impurities of the background electrolyte (`chemical noise') and air bubbles (Kuhn and Hoffstetter-Kuhn 1993; Xu et al. 1997). The unsatisfactory sample injection could also contribute to the low frequency noise due to the background buffer solution variations. The term *baseline drift* can refer to very low frequencies (Kuhn and Hoffstetter-Kuhn 1993) or to low frequencies in general (Solis et al. 2007). It should be mentioned that the detector type could also affect the baseline signal characteristics. For example, there are a range of detectors which can be used for CE or other techniques such as UV/Visible, fluorescence, electrochemical, conductivity, light scattering, mass spectral techniques. Some of the detectors measure bulk property of samples (e.g. conductivity and refractive index techniques) while some are for measuring solute properties (e.g. UV/Vis, fluorescence, electrochemical techniques). Bulk property detectors tend to have higher background signal which could vary due to background condition change. Solute detectors, on the other hand, usually have less background signal.

Since understanding the nature of the noise present in a system is important for an appropriate denoising of a signal (Perrin et al. 2001; Szymańska et al. 2007), numerous quantitative noise characterisation studies have been carried (Katsumine et al. 1999; Smith 2000; Smith 2007; Vaseghi 2008) to understand the spectral characteristics of the noise present in practical systems. The outcome of such studies may indicate the need for a unified noise model that does not partition the noise into low and high frequencies. For example, a brown noise model may be appropriate (Vaseghi 2008) or other models, e.g. a $1/f$ noise model (Katsumine et al. 1999; Smit and Walg 1975), or a correlated noise model (Perrin et al. 2001). The simplest method for noise modelling is to estimate the noise statistics from the signal-inactive periods (Vaseghi 2008). For instance, the noise in a system may be characterised by its power spectral density which can be estimated using a variety of methods (Cruz-Marcelo et al. 2008; Smith 2007). Of course, a noise process should really also be checked for stationarity by confirming its statistical parameters are constant over a sufficient interval of time. If a noise process is non-stationary, for example, the heteroscedastic noise (Li et al. 2006; Mittermayr et al. 1999; Mittermayr et al. 1997), then modelling techniques for time-varying stochastic processes will need to be employed (Vaseghi 2008). Where noise characterisation is impractical, adaptive filters or filters whose parameters can be adjusted or tuned manually in-situ may need to be employed for effective removal of noise.

It is worth mentioning that, while various models have been proposed for the low and high frequency noise, a unified approach is also required to handle the noise as a whole to simplify the signal processing process. The baseline drift and noise contributions could be modelled as interferences from different ends of frequency spectrum domain. Approaches may include Fourier and wavelet transforms. Some of these techniques will be reviewed in the next section

3. Signal processing techniques for peak extraction

3.1 Processing approaches

The aim of a peak extraction algorithm is to extract the separate peak components, $p_i(t)$, from an electropherogram. In general, when trying to identify the peaks in an unknown sample we aren't privy to information about when and where the different signal components will occur and in what measure. Only a measured signal is available and we have to solve an inverse problem (Mammone et al. 2007; Tarantola 2005) to identify the parameters in our model. Approaches to solving single-channel source separation problems must rely on additional information, with filtering, decomposition and grouping, and source modelling approaches having been used (Schmidt 2008). A typical signal processing strategy might consist of a number of steps including: (i) the removal/suppressing of the baseline noise, (ii) finding the peaks, and (iii) extracting and/or quantifying the peaks. Some approaches combine some of these steps into a single step. In the following sections we detail each of these steps in further detail.

3.2 Baseline noise removal

To remove the baseline noise from an electropherogram, it is often convenient to assume the noise components are confined within certain spectral ranges. Linear filters can then be used to suppress the content in those ranges (Horlick 1972; Rabiner et al. 1975). There are many standard filters that can be used to do this, some of which have been listed (Yang et al. 2009). In their review on peak detection algorithms for mass spectrometry, various open source software packages were compared and the filters employed were identified.

3.2.1 High frequency noise removal

For removing the high frequency baseline noise in analytical chemistry data, the moving average filter is a popular choice. Perhaps the most intuitive filter to understand, this Finite Impulse Response (FIR) filter (non-recursive filter) performs local averaging to attenuate the rapidly fluctuating components of a signal (Lyons 2004; Oppenheim et al. 2007). It is used for smoothing in the software PROcess (Li et al. 2005; Yang et al. 2009) for smoothing. However, with this filter, peaks tend to be flattened out. A widely cited seminal paper by Savitzky and Golay (Savitzky and Golay 1964) provides the details for a popular alternative for filtering out high frequency baseline noise. The filter, known as the Savitzky-Golay (SG) filter, has filter coefficients that implement least squares polynomial smoothing and can denoise signals and calculate derivatives with reduced peak degradation (Leptos et al. 2006; Peters et al. 2007; Vivó-Truyols and Schoenmakers 2006; Vivó-Truyols et al. 2005a). The filter has been extended and improved in many different ways (Browne et al. 2007). Other digital filters have also been used including the Kaiser filter (Mantini et al. 2007) and Gaussian related filters (Leptos et al. 2006; Yang et al. 2009). In general, digital filters have different time and frequency domain characteristics that are appropriate for different situations and often have parameters that must be optimized so that peak distortion is minimised (Hamming 1983) and the maximal amount of noise removed. It is possible to custom design a digital filter based on an ideal transfer function which, for low-pass (high-frequency attenuating) filter designs, can entail choosing an appropriate cut-off frequency (Lam and Isenhour 1981).

3.2.2 Low frequency noise removal

Digital linear filters can also be used for removing low frequency baseline noise (drift). However, the standard linear filtering approaches can have problems when the baseline noise is not confined to specific frequency bands or overlaps with the frequency bands of peaks. Similar problems arise in the processing of ECG signals, and are of principal concern since filtering can cause significant distortion to important key features (Mozaffary and Tinati 2005). Whilst non-linear filters may be one alternative strategy (Kiryu et al. 1994), the most common approach for filtering out the low frequency baseline noise in CE signals (and signals in related areas) is to estimate the low-frequency baseline and then subtract the estimate from the signal. Windowing and interpolation based methods are commonly used. For example, a moving window method, retaining minimums, was used to estimate the low frequency baseline noise component in chromatograms (Quéméner et al. 2007). The estimated component was then subtracted from the original signal to give a corrected chromatogram. The method appeared successful but required the width of the moving window to be set experimentally. A similar strategy was employed for baseline correction of CE data (Coombes et al. 2005; Szymańska et al. 2007). In another published work (Gras et al. 1999), selected values from windows were interpolated using cubic splines to estimate the low frequency baseline noise in MALDI-TOF mass spectra. A different strategy was employed with the signal trend being estimated by removing the peaks from a mass spectrum signal (Mantini et al. 2007). Other researchers have also applied curve fitting techniques (Coombes et al. 2005; Bernabé-Zafón et al. 2005; Gillies et al. 2006; Mazet et al. 2005).

3.2.3 Wavelet transformation for noise removal

Despite some successful demonstrations of (low and high frequency) baseline noise removal techniques, no single approach has proved so undeniably successful across various conditions that it has been universally adopted. Recently, however, there has been some interest in the use of relatively new signal processing techniques such as wavelets.

Wavelets (Burrus et al. 1998; Daubechies 1988; Grossmann and Morlet 1984; Mallat 1989, 1999) allow the simultaneous analysis of a signal's time and frequency (or scale) properties, which is particularly useful for the processing of transient signals. The continuous wavelet transform (CWT) of a signal, $f(t) \in L^2 R$, at a scale, s, and time u, is given by (Mallat 1999),

$$Wf(u,s) = \int_{-\infty}^{\infty} f(t) \frac{1}{\sqrt{s}} \psi^* \left(\frac{t-u}{s} \right) dt \qquad (4)$$

The "mother wavelet", $\psi(t) \in L^2 R$, is a function with zero average and "is well localised in both time and frequency" (Cohen and Kovačević 1996). Under certain conditions an inverse continuous wavelet transform also exists that allows the reconstruction of $f(t)$ (e.g. Burrus et al. 1998; Cohen and Kovačević 1996). The coefficients from the wavelet transformation provide an indication of the signal energy contained at various scales over time. It is therefore possible to calculate the energy density (scalogram) of a signal which can be used in signal analysis and/or graphically depicted in a time-frequency heat map.

For a discrete wavelet transform (DWT), the wavelet coefficients can be computed for a discrete grid of points $(s,u)_{n \in Z}$ (Cohen and Kovačević 1996). Practically though, a multiresolution analysis (MRA) (Mallat 1989) is most often used to give a series expansion for $f(t) \in L^2 R$, in terms of a scaling function, $\varphi(t)$, and wavelets, $\psi_{j,k}$ (Burrus et al. 1998),

$$f(t) = \sum_{k=-\infty}^{\infty} c(k)\varphi_k(t) + \sum_{j=0}^{\infty} \sum_{k=-\infty}^{\infty} d(j,k)\psi_{j,k}(t) \qquad (5)$$

where orthogonality constraints are placed on the expansion functions which form a basis. The first summation in Eq. (5) provides a low resolution approximation to $f(t)$, and each successive index, *j*, in the second summation adds additional detail to the signal. A fast DWT algorithm that employs the use of filter banks has been developed by Mallat for the calculation of the approximation coefficients *c(k)* and detail coefficients *d(j,k)* (Mallat 1989), and is frequently used in practice. Once the coefficients have been obtained for a signal, they are thresholded or processed in various ways, before the reconstruction part of Mallat's algorithm can be used to reconstruct the signal.

One of the most widely used applications of wavelets is in denoising. Donoho and Johnstone developed a wavelet denoising technique based on the thresholding of coefficients (Donoho 1995; Donoho and Johnstone 1994). The denoising works on the premise that the underlying signal can have a sparse representation where it is approximated by a small number of relatively large-amplitude coefficients, whereas noise will have its energy spread across a large number of small-amplitude coefficients (Burrus et al. 1998; Mallat 1999). Hence, thresholding can remove noise whilst keeping the underlying signal largely intact. Different thresholding strategies such as hard thresholding and soft thresholding can be applied (Burrus et al. 1998; Donoho 1995).

A comprehensive study on wavelet denoising in CE using thresholding based methods was recently performed (Perrin et al. 2001). In the study, a number of wavelets from different wavelet families were tested, with the Haar wavelet found to perform the best. High frequency noise was removed by filtering the detail coefficients using hard thresholding and the low frequency baseline noise was removed by filtering the approximation coefficients using soft thresholding. Soft thresholding was used so as to reduce peak distortion. After thresholding, the inverse wavelet transform was calculated to reconstruct the signal with impressive results. The strategy was developed to accommodate larger baseline drifts (Liu et al. 2003). Fig. 3 shows an example of signal denoising using wavelet transform. This line of work was further developed by implementing spatially adaptive thresholding for the denoising of DNA capillary electrophoresis signals (Wang and Gao 2008). Wavelet denoising strategies were also used (Ceballos et al. 2008) on pattern recognition in CE. Similar wavelet based denoising strategies to those cited above have also been applied to liquid chromatography data (Barclay and Bonner 1997; Shao et al. 2004), Raman spectroscopy (Hu et al. 2007), mass spectrometry data (Barclay and Bonner 1997; Coombes et al. 2005), as well as numerous other areas of research (Jagtiani et al. 2008; Komsta 2009).

The DWT is sufficient in many scenarios when removing high and low frequency noise from signals. However, in some situations it may be appropriate to use the continuous wavelet transform so that finer grained control over the scales used can be gained. In particular, Jakubowska has done some interesting work using the CWT for removing high and low frequency noise in voltammetry signals (Jakubowska 2008). After finding the CWT, certain scale bands were identified as containing noise, and these were excluded in the reconstruction of a signal using the inverse CWT. The method was also suitable for resolving overlapping peaks which will be discussed in the next sections.

3.3 Peak detection

Once the noise has been filtered, the next task is to find the peaks from the filtered electropherogram $v_f(t) = v(t) - n(t)$. Many different criteria can be used to determine whether a point corresponds to the apex of a peak (Yang et al. 2009). One simple strategy for finding the location of peaks in $v_f(t)$ might be to find those points that are above a threshold and/or correspond to a local maximum found by looking for positive to negative zero crossings in

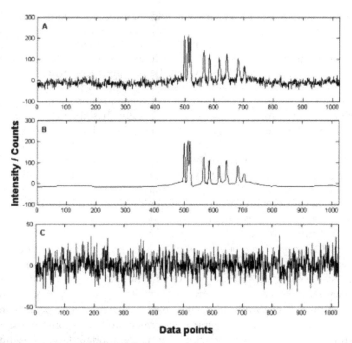

Fig. 3. An example of signal denoising by a discrete wavelet transform technique. (A)Original signal; (B) Denoised signal by db5 at decomposition level; (C) Removed noise. Reproduced from Liu et al. (2003) with permission from Wiley-VCH Verlag GmbH & Co. KGaA.

the signal derivative (or looking at just the second derivative). Thresholding and derivative based peak detection strategies are popular in various areas outside of CE. For example, an auto-threshold based peak detection algorithm for analyzing electrocardiograms was developed (Jacobson 2001); peaks in mass spectrometry data were found by selecting local maximum points with a signal-to-noise ratio above a certain value (Coombes et al. 2005; Mantini et al. 2007), and points of local maximum with an intensity sufficiently greater than neighbouring points were classified as peaks (Yasui et al. 2003). Zero-crossings of the derivative of signal were used to find the peaks in the analysis of partial discharge amplitude distributions (Carrato and Contin 1994). Derivatives of a smoothed signal are often used for peak detection in chromatography (Poole 2003). Signal peaks were detected in chromatograms after the Savitzky-Golay (SG) method was used to calculate derivatives (Peters et al. 2007). For CE, it was detailed how peaks can be identified through the location of inflection points (calculated using the second forward difference) (Graves-Morris et al. 2006).

Whilst derivative and threshold based techniques may be appropriate for some signals, low frequency baseline noise remaining in an electropherogram signal can disrupt threshold based peak detection (Wee et al. 2008), and high frequency noise can also cause problems especially for derivative based techniques, as discussed in (Lu et al. 2006). Recently, other approaches have been developed that are less susceptible to the inevitable unsuppressed noise that remains in the filtered electropherogram signal.

It is well established that wavelets can be used to detect singularities (points not differentiable) and edges (Mallat 1999; Struzik 2000). Singularities are detected by following across scales the local maxima of the wavelet transform. Using related multiscale zooming procedures, it is possible to detect discrete-time peaks even though the corresponding peaks may be continuous and differentiable. This idea has been applied to mass spectrometry data (Du et al. 2006). The CWT was firstly evaluated at different scales using the Mexican Hat wavelet to give a matrix of coefficients. Patterns evident in the matrix of coefficients were then analysed, with local maxima at each scale being linked across scales to give "ridges". Ridges meeting certain conditions were then used to indicate the location of peaks. The algorithm was later extended to peak detection in CE (Petkovic-Duran et al. 2008; Stewart et al. 2008). An example of peak detection using the technique is shown in Fig. 4, with comparison with three other techniques (Du et al. 2006; Mantini et al. 2007; Morris et al. 2005). Similar peak detection approaches have also been adopted in other areas, such as detecting evoked potentials (EPs) of the brain in electroencephalograms (EEGs) using the DWT (McCooey et al. 2005; McCooey and Kumar 2007).

A potential benefit of these wavelet approaches is that the pre-processing step of removing low and high frequency baseline noise is unnecessary, as the process of selecting ridge lines can account for the presence of noise. Reconstruction of signals based on the local maxima of the wavelet-transform modulus is also possible (Mallat and Hwang 1992), so it would be interesting to investigate whether reconstruction of peaks in CE could be performed in this way. Such a technique has already been applied for extracting EPs of the brain from EEGs (Zhang and Zhen 1997) in addition to detection of EPs (mentioned earlier).

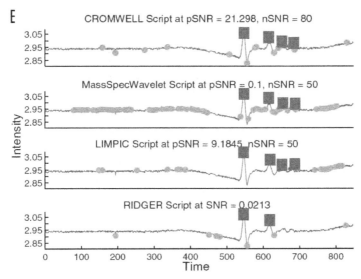

Fig. 4. An example of peak detection using a continuous wavelet transform technique and its comparison with other techniques. The techniques shown are, from top to bottom, CROMWELL (Morris et al. 2005), MassSpecWavelet Script (Du et al. 2006), LIMPIC (Mantini et al. 2007) and Ridger (Wee et al. 2008), respectively. Reproduced from (Wee et al. 2008) with permission from Wiley-VCH Verlag GmbH & Co. KGaA.

Whilst it is beyond the scope of this paper, it should be noted that when multiple data sets are available (i.e. data from multiple trials) statistical peak alignment or finding techniques can be applied (Ceballos et al. 2008; Coombes et al. 2005; Cruz-Marcelo et al. 2008; Dixon et al. 2006; Liu et al. 2008; Morris et al. 2005; Yu et al. 2008).

3.4 Peak resolution, peak extraction and quantification

After the baseline noise in the signal is suppressed and the peak locations are identified, peak components can be readily extracted by taking the portion of the signal above the constant baseline (B in Eq. 1), or $B + \delta$ where δ can account for a threshold that is above residual noise or tailing peaks) around the identified peak locations. A peak model can be fitted to the extracted peaks and measure important peak parameters such as area, height, skewness and so on. However, a single model may be insufficient and the curve fitting process is complicated when there are overlapping peaks that are not baseline resolved (Dyson 1998).

Usually in developing a CE method, the experimental conditions are optimised so as to ensure that all sample components are separated (Hanrahan et al. 2008; Vera-Candioti et al. 2008). However, such optimisation may not always be practicable and/or complete separation may be difficult, if not impossible, to achieve (Mammone et al. 2007; Sentellas et al. 2001; Zhang et al. 2007). As a result, unresolved peaks may be present in an electropherogram and this means quantitative measurements made directly on the peaks may be inaccurate (Dyson 1998). There is thus great need for signal processing techniques that are able to separate overlapping peaks especially in situations where conducting additional trials would be undesirable, infeasible or ineffectual.

Were the data acquired multi-dimensional (e.g. from multiple identical or dissimilar detectors or resulting from multiple trials under similar or dissimilar conditions), numerous statistical techniques would be at the researchers disposal to try and resolve overlapping peaks (Bocaz-Beneventi et al. 2002; Li et al. 2006; Sentellas et al. 2001; Zhang et al. 2007; Zhang and Li 2006). However, different approaches are necessary when the data from only a one-dimensional data vector (that results in a single electropherogram trace) is available. There are two main signal processing approaches that can be followed to analyse overlapping peaks.

The first approach is to try and extract the peak components from the signal using curve fitting. An accurate peak model is required and, if the number of peaks is known, the model can then be fitted to the peaks that are overlapping using non-linear least squares curve fitting techniques. Such techniques have been applied to CE (Jarméus and Emmer 2008; Vera-Candioti et al. 2008), as well as chromatography (Dasgupta 2008; Jin et al. 2008; Vivó-Truyols et al. 2005a, b), voltammetry (Huang et al. 1995) and gel electrophoresis (Kitazoe et al. 1983; Shadle et al. 1997) to name but a few. Fig. 5 shows an example of deconvolution technique (Vivó-Truyols et al. 2005b) for extracting overlapping peaks from a chromatographic signal. Such technique is also applicable for CE and other similar signals. However, the above-mentioned methods require a good peak model and initial parameter estimates must be sufficiently accurate for convergence of the curve fitting routines (Jarméus and Emmer 2008; Olazábal et al. 2004). In addition, in some cases the methods may be sensitive to noise and it may be difficult to automate the whole process (Jarméus and Emmer 2008; Vivó-Truyols et al. 2005a; Zhang et al. 2000). If there is poor separation, quantitative measurements made on the peaks may be inaccurate (Du et al. 2006).

The second approach to analysing overlapping peaks is to apply signal processing to increase the resolution of the peaks by decreasing their width but preserving their height or area. For example, voltammetric or polarographic peaks were resolved using a deconvolution method (Engblom 1990). In the technique, signals were transformed to the Fourier domain where the width of the peak (in the time domain) could be modified by dividing/multiplying the Fourier transform by a suitable function (note: multiplication in the Fourier domain corresponds to convolution in the time domain). By an appropriate function choice, peaks can be resolved in the time domain by having their widths reduced but their height or area preserved. A deconvolution approach was also detailed in Kauppinen et al. 1981 and more recently deconvolution/convolution approaches have been used in combination with wavelets (Wang et al. 2004; Zhang et al. 2000; Zhang et al. 2001; Zheng et al. 2000).

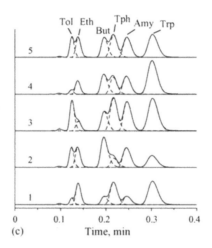

Fig. 5. An example of deconvolution of multi-overlapped chromatographic signal peaks. The five runs correspond to five mixtures eluted with 80% (m/m) methanol. The compounds are toluene (Tol), ethylbenzene (Eth), butylbenzene (But), o-terphenyl (Tph), amylbenzene (Amy), and triphenylene (Trp), respectively. Reprinted from Vivó-Truyols et al. (2005b) with permission from Elservier.

Perhaps the most popular approaches for dealing with overlapping peaks involve the use of wavelet transforms. These can be used to transform a signal with overlapping peaks into a new waveform with resolved peaks, where, in so doing, some peak parameters (e.g. area and location) can be preserved. Such approaches are suitable where quantification of certain peak parameters is of interest but the complete peak waveform is not required. For example, the discrete wavelet transform (DWT) was used (Shao et al. 1997) to process chromatograms by decomposing them into detail coefficients at different decomposition levels. The overlapping peaks for the chromatograms analysed had peaks in the detail coefficients that were well resolved at a specific level of decomposition. After baseline correction (which was needed as a secondary step) the areas under the peaks were used to provide an accurate quantitative measure of the concentration.

The continuous wavelet transform (CWT) has also been used to resolve overlapping peaks in CE. Unlike the dyadic levels of decomposition typically used with the DWT's orthogonal wavelets, in non-orthogonal wavelet analysis the choice of scales is arbitrary (Torrence and

Compo 1998). For example, the maximum wavelet scale was chosen to resolve coefficient peaks Jiao et al. 2008. At the selected scale, the resolved coefficient peaks could then be quantitatively analysed. Other researchers have applied similar approaches using the CWT (Jakubowska and Kubiak 2008; Shao et al. 1997; Shao and Sun 2001; Wang et al. 2004). In choosing the scale, the aim is to minimise the noise but ensure peaks are baseline resolved, and the mother wavelet chosen as well as the signals being analysed impact upon the results. Where resolution at specific scales in the wavelet coefficients is not evident, information across scales can also been utilised (Xiao-Quan et al. 1999) as it is for peak finding (see Section 4.3).

In difficult problems, peaks may still not be resolved by the CWT coefficients. In such cases it may be useful or necessary to select a range of scales and reconstruct a time-domain signal using the inverse CWT (Jakubowska 2008). An appropriate choice for the scale band used can then lead to resolution of peaks in the reconstructed signal (Torrence and Compo 1998).

Whilst it is clear that a range of techniques exist for extracting quantitative information about overlapping peaks, there is still a need for techniques that are able to extract the peak components in their entirety. The process of resolving peaks with wavelet transforms does not generally permit the peak components to be extracted directly, as usually some of the peak parameters are modified during transformation. Even though this isn't a problem for obtaining some quantitative information, it may be useful to have a complete representation for the peaks. In addition, most of the techniques demonstrated are for particular problems or setups. Generalised approaches to resolving overlapping peaks are needed that make minimal assumptions about the peak components and baseline noise.

4. Assessing algorithm and system performance

As this review has revealed, there are numerous approaches available for processing signals in capillary electrophoresis, each with a different signal processing strategy and demonstration on synthetic and/or real test data. However, little effort has been devoted to the comparison of performance from these different methods in spite of the few published work (Barclay and Bonner 1997; Cruz-Marcelo et al. 2008; Wee et al. 2008; Yang et al. 2009). In order to select or develop most efficient method, comparative studies are needed to test the methods on the same data (Leptos et al. 2006). One way to enable this happening is to make the software algorithms free and publically available so that other researchers can use and compare. One good example is that of Mantini and co-workers who provide both the algorithm source code and test data as additional material accompanying their paper (Mantini et al. 2007). Perhaps the more practical way is to standardise the test data sets so that different algorithms can be compared using the same data for their performance in key parameters such as noise removal, peak detection, peak resolution, peak extraction and quantification, and computing efficient, speed and power requirements. This can be viewed as a step towards reproducible research (Arora and Barak 2009; Deb 2001; Zheng et al. 1998). This type of approach is common in other areas of research. For example, in the area of multi-objective evolutionary optimisation (MOEA), there are standard test problems that algorithms are compared against (Deb 2001). Since that algorithm performance is likely to be dataset dependent, there should be a range of standardised benchmark datasets which could be obtained from real experimental setups or generated synthetically, and made publically available.

Standard performance indicators are needed in addition to standardised test data, so that algorithms can be quantitatively assessed and compared. For example, peak detection algorithms could be assessed by their false discovery rate (FDR) and sensitivity (Cruz-Marcelo et al. 2008; Wee et al. 2008; Yang et al. 2009) or receiver operating characteristic (ROC) curve (Mantini et al. 2007). For assessing noise removal efficiency or for evaluating the preservation of peak properties after peak resolution, different measures that might be used include: root square error (RSE) or integrated square error (ISE) (Jagtiani et al. 2008), root mean square (RMS) (Barclay and Bonner 1997), relative error (RE) (Zhang et al. 2001; Zheng et al. 1998), individual sum of squared residuals (Vivó-Truyols et al. 2005b), signal-to-noise ratio (SNR) and correlation coefficient (Jakubowska and Kubiak 2008). In addition to these performance indicators, an analysis of an algorithm's computational complexity (Arora and Barak 2009) would also be worth reporting. If this approach were widely adopted, then the performance of algorithms could be readily compared.

Assessing algorithm performance is one step towards complete system characterisation and performance assurances. With appropriate quantification of the noise processes present in a system and bounded variation in the peak model, in conjunction with knowledge regarding the performance of the signal processing algorithms used, a researcher should be able to specify the quantitative properties of peaks along with a determined uncertainty measure. This would also allow systems to be compared with other (including commercial) systems and would ensure objective assessment of results could be made.

5. Pattern matching - inferring chemical identity

Given the successful extraction of peaks from an electropherogram, the next step is to use this summary information to identify the present chemicals. Substantial effort has been devoted to the algorithm development for such a purpose (Fruetel et al. 2006; García-Péreza et al. 2008; Liu et al. 2008; Stein and Scott 1994). As this process requires reference to a library of peak information, extracted from known chemicals measured in similar environment, we will refer to it as *Pattern Matching*.

Due to uncertain inputs, the pattern matching is an inference process, resulting in probabilities. Inferring about the general composition of a sample, without reference to particular list of chemicals, is beyond what we consider possible from electrophoresis, as the solution for this question, based on electrophoresis results, is materially under-constraint. In general, we attempt to answer the following question:

What are the odds that a particular chemical (i) is present in the sample, given the peaks expected from chemical i and other chemicals in the same environment?

$$\frac{P(Chemical_i \ / \ DI\ )}{P(\overline{Chemical_i} \ / \ DI\ )} = ? \ , \qquad (6)$$

where D is the extracted peaks data, and I is the background library information.

The pattern matching is encumbered by several real life constrains: the accuracy of peaks extraction is not assured; It is not certain that the peaks extracted from two measurements, performed on same sample and in the same environment, will be the same. The list of known chemicals may not be exhaustive; chemicals for which we do not have peaks information (library) may be present in the sample. It is possible that the measurement environment is not identical to the environment in which the library was created, leading to different peaks extracted for same sample. It is also possible that several other chemicals will be present in the sample, influencing peaks extracted and attributed to the chemical of interest.

An effective pattern matching will need to reliably account for these uncertainties, and consistently indicate the probabilities of chemicals presence, using our understanding that several other known and unknown chemicals may be present in the sample. This, in general, cannot be solely done using naive matching of the extracted peaks to the library peaks of the chemical of interest, as it ignores the possible presence of other chemicals. And, comparison of separate calculation of likelihoods of the peaks measured for each of the known chemicals ignores other information in hand, such as the possibility of presence of unknown chemicals and concurrent presence of chemicals. The pattern matching process will also benefit from ability to mount the evidence, or in other words, to learn: given peaks measured in one environment, the adequate process should combine this information with information extracted from measurement of the same sample in a different environment. Such a pattern matching process for electrophoresis is, to the best of our knowledge, not defined yet, and is the subject of future work.

6. Concluding remarks

In this paper we have provided an overview of the signal processing methods that have been used in capillary electrophoresis and other related areas. We firstly discuss the various models proposed in the literature for modelling peak shapes and baseline noise. Signal processing techniques for extracting peaks from the signal are then reviewed. This covers noise removal such as digital filters and wavelet transforms, peak detection, peak extraction and quantification. We also discuss possible approaches for assessing algorithm development and system performance.

The problem of identifying peaks could be regarded as feature extraction, and problems of such a nature are not confined to analytical chemistry. There exists significant opportunity to apply ideas and adapt techniques from other disciplines (such as pattern recognition, adaptive control, particle swarm optimisation, evolutionary multi-objective optimisation, machine learning, artificial neural networks (ANN) and artificial intelligence generally), to the processing of signals in CE. Indeed, for real systems that may have drifting parameters or characteristics, techniques from other disciplines may be necessary to produce adaptable signal processing algorithms.

With benchmark testing and performance assessment, we should be able to develop algorithms to realise effective and efficient automatic signal analysis and quantification. Peaks quantified could then be used as the input to an inference pattern matching stage to determine the analyte components present in a sample. These steps will help facilitate the development of CE systems that reliably perform to specification and allow users to focus on the experimental results of separation.

7. Acknowledgments

Financial support from the Australian Government Department of the Prime Minister and Cabinet through the National Security Science and Technology Branch, the Australian Federal Police (AFP) and the Australian Customs Service (ACS) is greatly acknowledged. The authors would also like to thank Ms. Karolina Petkovic-Duran for providing the data used in Fig. 1.

8. References

Arora S, Barak B (2009) Computational Complexity: A Modern Approach. Cambridge University Press, Cambridge

Barclay VJ, Bonner RF (1997) Application of Wavelet Transforms to Experimental Spectra: Smoothing, Denoising, and Data Set Compression. Anal Chem 69:78-90

Beckers JL, Everaerts FM (1997) System peaks in capillary zone electrophoresis What are they and where are they coming from? J Chromatogr A 787:235-242

Bernabé-Zafón V, Torres-Lapasió JR, Ortega-Gadea S, Simó-Alfonso EF, Ramis-Ramos G (2005) Capillary electrophoresis enhanced by automatic two-way background correction using cubic smoothing splines and multivariate data analysis applied to the characterisation of mixtures of surfactants. J Chromatogr A 1065:301-313

Bocaz-Beneventi G, Latorre R, Farková M, Havel J (2002) Artificial neural networks for quantification in unresolved capillary electrophoresis peaks. Anal Chim Acta 452:47-63

Browne M, Mayer N, Cutmore TRH (2007) A multiscale polynomial filter for adaptive smoothing. Digital Signal Process 17:69-75

Burrus CS, Gopinath RA, Guo H (1998) Introduction to Wavelets and Wavelet Transforms: A Primer. Prentice-Hall, Upper Saddle River

Carrato S, Contin A (1994) Application of a peak detection algorithm for the shape analysis of partial discharges amplitude distributions. Conference Record of the 1994 IEEE International Symposium on Electrical Insulation:288-291

Ceballos GA, Paredes JL, Hernández LF (2008) Pattern recognition in capillary electrophoresis data using dynamic programming in the wavelet domain. Electrophoresis 29:2828-2840

Cohen A, Kovačević J (1996) Wavelets: the mathematical background. Proc IEEE 84:514-522

Coombes KR, Tsavachidis S, Morris JS, Baggerly KA, Hung M-C, Kuerer HM (2005) Improved peak detection and quantification of mass spectrometry data acquired from surface-enhanced laser desorption and ionization by denoising spectra with the undecimated discrete wavelet transform. Proteomics 5:4107-4117

Couch LW, II (1990) Digital and Analog Communication Systems. 4th edn. Macmillan, New York

Cruz-Marcelo A, Guerra R, Vannucci M, Li Y, Lau CC, Man T-K (2008) Comparison of algorithms for pre-processing of SELDI-TOF mass spectrometry data. Bioinformatics 24:2129-2136

Dasgupta PK (2008) Chromatographic peak resolution using Microsoft Excel Solver: The merit of time shifting input arrays. J Chromatogr A 1213:50-55

Daubechies I (1988) Orthonormal bases of compactly supported wavelets. Commun Pure Appl Math 41:909-996

Deb K (2001) Multi-Objective Optimization using Evolutionary Algorithms. John Wiley and Sons, Chichester

Dixon SJ, Brereton RG, Soini HA, Novotny MV, Penn DJ (2006) An automated method for peak detection and matching in large gas chromatography-mass spectrometry data sets. J Chemom 20:325-340

Donoho DL (1995) De-noising by soft-thresholding. IEEE Trans Inf Theory 41:613-627

Donoho DL, Johnstone IM (1994) Ideal spatial adaptation by wavelet shrinkage. Biometrika 81:425-455

Du P, Kibbe WA, Lin SM (2006) Improved peak detection in mass spectrum by incorporating continuous wavelet transform-based pattern matching. Bioinformatics 22:2059-2065

Dyson N (1998) Chromatographic Integration Methods. Royal Society of Chemistry Chromatography Monographs, 2nd edn. Royal Society of Chemistry, Cambridge

Engblom SO (1990) The Fourier transform of a voltammetric peak and its use in resolution enhancement. J Electroanal Chem 296:371-394

Fruetel JA, West JAA, Debusschere BJ, Hukari K, Lane TW, Najm HN, Jose Ortega J, Renzi RF, Shokair I, VanderNoot VA (2006) Identification of Viruses Using Microfluidic Protein Profiling and Bayesian Classification. Anal Chem 80:9006-9012

García-Alvarez-Coque MC, Simó-Alfonso EF, Sanchis-Mallols JM, Baeza-Baeza JJ (2005) A new mathematical function for describing electrophoretic peaks. Electrophoresis 26:2076-2085

García-Péreza I, Vallejo M, García A, Legido-Quigley C, Barbas C (2008) Metabolic fingerprinting with capillary electrophoresis. J Chromatogr A 1204:130-139

Gaš B, Hruška V, Dittmann M, Bek F, Witt K (2007) Prediction and understanding system peaks in capillary zone electrophoresis. J Sep Sci 30:1435-1445

Gebauer P, Boček P (1997) System peaks in capillary zone electrophoresis I. Simple model of vacancy electrophoresis. J Chromatogr A 772:73-79

Gillies P, Marshall I, Asplund M, Winkler P, Higinbotham J (2006) Quantification of MRS data in the frequency domain using a wavelet filter, an approximated Voigt lineshape model and prior knowledge. NMR Biomed 19:617-626

Gras R, Müller M, Gasteiger E, Gay S, Binz P-A, Bienvenut W, Hoogland C, Sanchez J-C, Bairoch A, Hochstrasser DF, Appel RD (1999) Improving protein identification from peptide mass fingerprinting through a parameterized multi-level scoring algorithm and an optimized peak detection. Electrophoresis 20:3535-3550

Graves-Morris PR, Fell AF, Bensalem M (2006) Parameterisation of symmetrical peaks in capillary electrophoresis using [3/2]-type rational approximates. J Comput Appl Math 189:220-227

Grossman PD, Colburn JC (eds) (1992) Capillary Electrophoresis: Theory and Practice. Academic Press, San Diego

Grossmann A, Morlet J (1984) Decomposition of Hardy functions into square integrable wavelets of constant shape. SIAM J Math Anal 15:723-736

Grushka E (1972) Characterization of Exponentially Modified Gaussian Peaks in Chromatography. Anal Chem 44:1733-1738

Guijt RM, Evenhuis CJ, Macka M, Haddad PR (2004) Conductivity detection for conventional and miniaturised capillary electrophoresis systems. Electrophoresis 25:4032-4057

Hamming RW (1983) Digital Filters. Prentice-Hall, Englewood Cliffs

Hanrahan G, Montes R, Gomez FA (2008) Chemometric experimental design based optimization techniques in capillary electrophoresis: a critical review of modern applications. Anal Bioanal Chem 390:169-179

Horlick G (1972) Digital data handling of spectra utilizing Fourier transformations. Anal Chem 44:943-947

Hu Y, Jiang T, Shen A, Li W, Wang X, Hu J (2007) A background elimination method based on wavelet transform for Raman spectra. Chemom Intell Lab Syst 85:94-101

Huang W, Henderson TLE, Bond AM, Oldham KB (1995) Curve fitting to resolve overlapping voltammetric peaks: model and examples. Anal Chim Acta 304:1-15

Inczédy J, Lengyel T, Ure AM (eds) (1998) Compendium of Analytical Nomenclature: Definitive Rules 1997. 3rd edn. Blackwell Science, Oxford

Issaq HJ (2001) The role of separation science in proteomics research. Electrophoresis 22:3629-3638

Jacobsen NE (2007) NMR Spectroscopy Explained: Simplified Theory, Applications and Examples for Organic Chemistry and Structural Biology. John Wiley and Sons, Hoboken

Jacobson ML (2001) Auto-threshold peak detection in physiological signals. Engineering in Medicine and Biology Society, 2001. Proceedings of the 23rd Annual International Conference of the IEEE.

Jagtiani AV, Sawant R, Carletta J, Zhe J (2008) Wavelet transform-based methods for denoising of Coulter counter signals. Meas Sci Technol 19:065102

Jakubowska M (2008) Inverse continuous wavelet transform in voltammetry. Chemom Intell Lab Syst 94:131-139

Jakubowska M, Kubiak WW (2008) Signal processing in normal pulse voltammetry by means of dedicated mother wavelet. Electroanalysis 20:185-193

Jarméus A, Emmer Å (2008) CE Determination of monosaccharides in pulp using indirect detection and curve-fitting. Chromatographia 67:151-155

Jiao L, Gao S, Zhang F, Li H (2008) Quantification of components in overlapping peaks from capillary electrophoresis by using continues [sic] wavelet transform method. Talanta 75:1061-1067

Jin G, Xue X, Zhang F, Zhang X, Xu Q, Jin Y, Liang X (2008) Prediction of retention times and peak shape parameters of unknown compounds in traditional Chinese medicine under gradient conditions by ultra performance liquid chromatography. Anal Chim Acta 628:95-103

Johansson G, Isaksson R, Harang V (2003) Migration time and peak area artifacts caused by systemic effects in voltage controlled capillary electrophoresis. J Chromatogr A 1004:91-98

Kappes T, Hauser PC (1999) Electrochemical detection methods in capillary electrophoresis and applications to inorganic species. J Chromatogr A 834:89-101

Katsumine M, Iwaki K, Matsuda R, Hayashi Y (1999) Routine check of baseline noise in ion chromatography. J Chromatogr A 833:97-104

Kauppinen JK, Moffatt DJ, Mantsch HH, Cameron DG (1981) Fourier transforms in the computation of self-deconvoluted and first-order derivative spectra of overlapped band contours. Anal Chem 53:1454-1457

Kiryu T, Kaneko H, Saitoh Y Artifact elimination using fuzzy rule based adaptive nonlinear filter. In: Acoustics, Speech, and Signal Processing, 1994. ICASSP-94., 1994 IEEE International Conference on, 19-22 Apr 1994 1994. pp III/613-III/616

Kitazoe Y, Miyahara M, Hiraoka N, Ueta H, Utsumi K (1983) Quantitative determination of overlapped proteins in sodium dodecyl sulfate-polyacrylamide gel electrophoresis. Anal Biochem 134:295-302

Komsta Ł (2009) Suppressing the charged coupled device noise in univariate thin-layer videoscans: A comparison of several algorithms. J Chromatogr A 1216:2548-2553

Kubáň P, Hauser PC (2004) Contactless conductivity detection in capillary electrophoresis: A review. Electroanalysis 16:2009-2001

Kubáň P, Hauser PC (2009) Ten years of axial capacitively coupled contactless conductivity detection for CZE - a review. Electrophoresis 30:176-188
Kuhn R, Hoffstetter-Kuhn S (1993) Capillary Electrophoresis: Principles and Practice. Springer-Verlag, Berlin
Lam RB, Isenhour TL (1981) Equivalent width criterion for determining frequency domain cutoffs in fourier transform smoothing. Anal Chem 53:1179-1182
Lan K, Jorgenson JW (2001) A hybrid of exponential and gaussian functions as a simple model of asymmetric chromatographic peaks. J Chromatogr A 915:1-13
Leptos KC, Sarracino DA, Jaffe JD, Krastins B, Church GM (2006) MapQuant: Open-source software for large-scale protein quantification. Proteomics 6:1770-1782
Li H, Hou J, Wang K, Zhang F (2006) Resolution of multicomponent overlapped peaks: A comparison of several curve resolution methods. Talanta 70:336-343
Li X, Gentleman R, Lu X, Shi Q, Iglehart JD, Harris L, Miron A (2005) SELDI-TOF mass spectrometry protein data. In: Gentleman R, Irizarry RA, Carey VJ, Dudoit S, Huber W (eds) Bioinformatics and Computational Biology Solutions Using R and Bioconductor. Springer, New York, pp 91-109
Liu B-F, Sera Y, Matsubara N, Otsuka K, Terabe S (2003) Signal denoising and baseline correction by discrete wavelet transform for microchip capillary electrophoresis. Electrophoresis 24:3260-3265
Liu J, Yu W, Wu B, Zhao H (2008) Bayesian mass spectra peak alignment from mass charge ratios. Cancer Inform 6:217-241
Lu W, Nystrom MM, Parikh PJ, Fooshee DR, Hubenschmidt JP, Bradley JD, Low DA (2006) A semi-automatic method for peak and valley detection in free-breathing respiratory waveforms. Med Phys 33:3634-3636
Lyons RG (2004) Understanding Digital Signal Processing. Prentice Hall, Upper Saddle River
Macka M, Haddad PR, Gebauer P, Boček P (1997) System peaks in capillary zone electrophoresis 3. Practical rules for predicting the existence of system peaks in capillary zone electrophoresis of anions using indirect spectrophotometric deetection. Electrophoresis 18:1998-2007
Mallat S, Hwang WL (1992) Singularity detection and processing with wavelets. IEEE Trans Inf Theory 38:617-643
Mallat SG (1989) A theory for multiresolution signal decomposition: The wavelet representation. IEEE Trans Pattern Anal Mach Intell 11:674-693
Mallat SG (1999) A Wavelet Tour of Signal Processing. 2nd edn. Academic Press, San Diego
Mammone N, Fiaschè M, Inuso G, Foresta FL, Morabito FC, Versaci M (2007) Information theoretic learning for inverse problem resolution in bio-electromagnetism. LNAI 4694 (4694):414-421
Mantini D, Petrucci F, Pieragostino D, Boccio PD, Nicola MD, Ilio CD, Federici G, Sacchetta P, Comani S, Urbani A (2007) LIMPIC: a computational method for the separation of protein MALDI-TOF-MS signals from noise. BMC Bioinf 8:101
Marzilli LA, Bedard P, Mabrouk PA (1997) Learning to Learn: An Introduction to Capillary Electrophoresis. Chem Educ 1 (6):1-12
Mazet V, Carteret C, Brie D, Idier J, Humbert B (2005) Background removal from spectra by designing and minimising a non-quadratic cost function. Chemom Intell Lab Syst 76:121-133

McCooey C, Kumar DK, Cosic I Decomposition of evoked potentials using peak detection and the discrete wavelet transform. In: Proceedings of the 2005 IEEE Engineering in Medicine and Biology 27th Annual Conference, 2005. pp 2071-2074

McCooey CG, Kumar D Automated peak decomposition of evoked potential signals using wavelet transform singularity detection. In: Proceedings of the 29th Annual International Conference of the IEEE Engineering in Medicine and Biology Society, 2007. pp 23-26

Mittermayr CR, Lendl B, Rosenberg E, Grasserbauer M (1999) The application of the wavelet power spectrum to detect and estimate 1/f noise in the presence of analytical signals. Anal Chim Acta 388:303-313

Mittermayr CR, Rosenberg E, Grasserbauer M (1997) Detection and estimation of heteroscedastic noise by means of the wavelet transform. Anal Commun 34:73-75

Morris JS, Coombes KR, Koomen J, Baggerly KA, Kobayashi R (2005) Feature extraction and quantification for mass spectrometry in biomedical applications using the mean spectrum. Bioinformatics 21 (9):1764-1775

Mozaffary B, Tinati MA (2005) ECG Baseline Wander Elimination using Wavelet Packets. World Acad Sci Eng Tech 3:14-16

Naish PJ, Hartwell S (1988) Exponentially Modified Gaussian Functions - a Good Model for Chromatographic Peaks in Isocratic HPLC? Chromatographia 26:285-296

Oda RP, Landers JP (1997) Introduction to Capillary Electrophoresis. In: Landers JP (ed) Handbook of Capillary Electrophoresis. CRC Press, Boca Raton, pp 1-47

Olazábal V, Prasad L, Stark P, Olivares JA (2004) Application of wavelet transforms and an approximate deconvolution method for the resolution of noise overlapped peaks in DNA capillary electrophoresis. Analyst 129:73-81

Oppenheim AV, Willsky AS, Young IT (2007) Signals and Systems. Prentice-Hall, Englewood Cliffs

Parris NA (1984) Instrumental Liquid Chromatography: A Practical Manual on High-Performance Liquid Chromatographic Methods. 2nd edn. Elsevier, Amsterdam

Perrin C, Walczak B, Massart DL (2001) The Use of Wavelets for Signal Denoising in Capillary Electrophoresis. Anal Chem 73:4903-4917

Peters S, Vivó-Truyols G, Marriott PJ, Schoenmakers PJ (2007) Development of an algorithm for peak detection in comprehensive two-dimensional chromatography. J Chromatogr A 1156:14-24

Petkovic-Duran K, Zhu Y, Chen C, Swallow A, Stewart R, Hoobin P, Leech P, Ovenden S Hand-Held Analyzer Based on Microchip Electrophoresis with Contactless Conductivity Detection for Measurement of Chemical Warfare Agent Degradation Products. In: Nicolau DV, Metcalfe G (eds) Biomedical Applications of Micro- and Nanoengineering IV and Complex Systems, 2008. Proceedings of SPIE Vol. 7270 (SPIE, Bellingham, WA), p 72700Q

Polesello S, Valsecchi SM (1999) Electrochemical detection in the capillary electrophoresis analysis of inorganic compounds. J Chromatogr A 834:103-116

Poole CF (2003) The Essence of Chromatography. Elsevier, Amsterdam

Poppe H (1999) System peaks and non-linearity in capillary electrophoresis and high-performance liquid chromatography. J Chromatogr A 831:105-121

Quéméner B, Bertrand D, Marty I, Causse M, Lahaye M (2007) Fast data preprocessing for chromatographic fingerprings of tomato cell wall polysaccharides using chemometric methods. J Chromatogr A 1141:41-49

Rabiner LR, Sambur MR, Schmidt CE (1975) Applications of a nonlinear smoothing algorithm to speech processing. IEEE Trans Acoust Speech Signal Process ASSP-23:552-557

Reichenbach SE, Carr PW, Stoll DR, Tao Q (2009) Smart Templates for peak pattern matching with comprehensive two-dimensional liquid chromatography. J Chromatogr A 1216:3458-3466

Savitzky A, Golay MJE (1964) Smoothing and Differentiation of Data by Simplified Least Squares Procedures. Anal Chem 36:1627-1639

Schmidt MN (2008) Single-channel source separation using non-negative matrix factorization (PhD thesis), Technical University of Denmark,

Sellmeyer H, Poppe H (2002) Position and intensity of system (eigen) peaks in capillary zone electrophoresis. J Chromatogr A 960:175-185

Sentellas S, Saurina J, Hernández-Cassou S, Galceran MT, Puignou L (2001) Resolution and quantification in poorly separated peaks from capillary zone electrophoresis using three-way data analysis methods. Anal Chim Acta 431:49-58

Shadle SE, Allen DF, Guo H, Pogozelski WK, Bashkin JS, Tullius TD (1997) Quantitative analysis of electrophoresis data: novel curve fitting methodology and its application to the determination of a protein-DNA binding constant. Nucleic Acids Res 25:850-860

Shao X, Cai W, Sun P, Zhang M, Zhao G (1997) Quantitative determination of the components in overlapping chromatographic peaks using wavelet transform. Anal Chem 69:1722-1725

Shao X, Sun L (2001) An application of the continuous wavelet transform to resolution of multicomponent overlapping analytical signals. Anal Lett 34 (2):267-280

Shao X, Wang G, Wang S, Su Q (2004) Extraction of mass spectra and chromatographic profiles from overlapping GC/MS signal with background. Anal Chem 76:5143-5148

Smit HC, Walg HL (1975) Base-Line noise and detection limits in signal-integrating analytical methods. Application to Chromatography. Chromatographia 8:311-323

Smith AD (ed) (2000) Oxford Dictionary of Biochemistry and Molecular Biology. Rev. ed. edn. Oxford University Press, Oxford

Smith JO, III (2007) Spectral Audio Signal Processing (March 2007 Draft)

Solis A, Rex M, Campiglia AD, Sojo P (2007) Accelerated multiple-pass moving average: A novel algorithm for baseline estimation in CE and its application to baseline correction on real-time bases. Electrophoresis 28:1181-1188

Stein SE, Scott DR (1994) Optimization and testing of mass spectral library search algorithms for compound identification. J Am Soc Mass Spectrum 5:859-866

Stewart R, Wee A, Grayden DB, Zhu Y Capillary Electrophoresis (CE) Peak Detection using a Wavelet Transform Technique. In: Nicolau DV, Metcalfe G (eds) Biomedical Applications of Micro- and Nanoengineering IV and Complex Systems, 2008. Proceedings of SPIE Vol. 7270 (SPIE, Bellingham, WA), p 727012

Struzik ZR (2000) Determining local singularity strengths and their spectra with the wavelet transform. Fractals 8:163-179

Szymańska E, Markuszewski MJ, Capron X, van Nederkassel A-M, Heyden YV, Markuszewski M, Krajka K, Kaliszan R (2007) Increasing conclusiveness of metabonomic studies of cheminformatic preprocessing of capillary electrophoretic data on urinary nucleoside profiles. J Pharm Biomed Anal 43:413-420

Tanyanyiwa J, Galliker B, Schwarz MA, Hauser PC (2002) Improved capacitively coupled conductivity detector for capillary electrophoresis. Analyst 127:214-218

Tarantola A (2005) Inverse Problem Theory and Methods for Model Parameter Estimation. Society for Industrial and Applied Mathematics, Philadelphia

Torrence C, Compo GP (1998) A Practical Guide to Wavelet Analysis. Bull Am Meteorol Soc 79:61-78

Vaseghi SV (2008) Advanced Digital Signal Processing and Noise Reduction. 4th edn. John Wiley and Sons, Chichester

Vera-Candioti L, Culzoni MJ, Olivieri AC, Goicoechea HC (2008) Chemometric resolution of fully overlapped CE peaks: Quantification of carbamazepine in human serum in the presence of several interferences. Electrophoresis 29:4527-4537

Vivó-Truyols G, Schoenmakers PJ (2006) Automatic selection of optimal Savitzky-Golay smoothing. Anal Chem 78:4598-4608

Vivó-Truyols G, Torres-Lapasió JR, van Nederkassel AM, Heyden YV, Massart DL (2005a) Automatic program for peak detection and deconvolution of multi-overlapped chromatographic signals: Part I: Peak detection. J Chromatogr A 1096:133-145

Vivó-Truyols G, Torres-Lapasió JR, van Nederkassel AM, Heyden YV, Massart DL (2005b) Automatic program for peak detection and deconvolution of multi-overlapped chromatographic signals: Part II: Peak model and deconvolution algorithms. J Chromatogr A 1096:146-155

Wang J (2005) Electrochemical detection for capillary electrophoresis microchips: A review. Electroanalysis 17:1133-1140

Wang Y, Gao Q (2008) Spatially adaptive stationary wavelet thresholding for the denoising of DNA capillary electrophoresis signal. J Anal Chem 63:768-774

Wang Y, Mo J, Chen X (2004) 2nd-order spline wavelet convolution method in resolving chemical overlapped peaks. Sci China, Ser B Chem 47:50-58

Wee A, Grayden DB, Zhu Y, Petkovic-Duran K, Smith D (2008) A Continuous Wavelet Transform Algorithm for Peak Detection. Electrophoresis 29:4215-4225

Xiao-Quan L, Xi-Wen W, Jin-Yuan M, Jing-Wan K, Jin-Zhang G (1999) Electroanalytical signal processing method based on B-spline wavelets analysis. Analyst 124:739-744

Xu X, Kok WT, Poppe H (1997) Noise and baseline disturbances in indirect UV detection in capillary electrophoresis. J Chromatogr A 786:333-345

Yang C, He Z, Yu W (2009) Comparison of public peak detection algorithms for {MALDI} mass spectrometry data analysis. BMC Bioinf 10:4

Yasui Y, Pepe M, Thompson ML, Adam B-L, Wright GL, Jr., Qu Y, Potter JD, Winget M, Thornquist M, Feng Z (2003) A data-analytic strategy for protein biomarker discovery: profiling of high-dimensional proteomic data for cancer detection. Biostatistics 4:449-463

Yu W, He Z, Liu J, Zhao H (2008) Improving mass spectrometry peak detection using multiple peak alignment results. J Proteome Res 7:123-129

Zemann AJ, Schnell E, Volgger D, Bonn GK (1998) Contactless conductivity detection for capillary electrophoresis. Anal Chem 70:563-567

Zhang F, Chen Y, Li H (2007) Application of Multivariate curve resolution-alternating least square methods on the resolution of overlapping CE peaks from different separation conditions. Electrophoresis 28:3674-3683

Zhang F, Li H (2006) Resolution of overlapping capillary electrophoresis peaks by using chemometric analysis: improved quantification by using internal standard. Chemom Intell Lab Syst 82:184-192

Zhang J, Zhen C (1997) Extracting evoked potentials with the singularity detection technique. IEEE Eng Med Biol Mag 16:155-161

Zhang XQ, Zheng JB, Gao H (2000) Comparison of wavelet transform and Fourier self-deconvolution (FSD) and wavelet FSD for curve fitting. Analyst 125:915-919

Zhang Y, Mo J, Xie T, Cai P, Zou X (2001) Application of spline wavelet self-convolution in processing capillary electrophoresis overlapped peaks with noise. Anal Chim Acta 437:151-156

Zheng J, Zhang H, Gao H (2000) Wavelet-Fourier self-deconvolution. Sci China, Ser B Chem 43:1-9

Zheng X-P, Mo J-Y, Cai P-X (1998) Simultaneous application of spline wavelet and Riemann-Liouville transform filtration in electroanalytical chemistry. Anal Commun 35:57-59

7

Understanding Tools and Techniques in Protein Structure Prediction

Geraldine Sandana Mala John[1], Chellan Rose[1] and Satoru Takeuchi[2] [1]*Central Leather Research Institute*
[2]*Factory of Takeuchi Nenshi ,Takenen*
[1]*India*
[2]*Japan*

1. Introduction

Protein structure prediction is an important area of protein science. Every protein has a primary structure, its sequence; a secondary structure, the helices and sheets; tertiary structure, the fold of the protein; and for some, the quaternary structure, multimeric formation of its polypeptide subunits. Protein structure has been experimented for the past several decades by physical and chemical methods. The dawn of protein sequencing began early in 1950s upon complete sequencing of insulin and then, ribonuclease. A key step towards the rapid increase in the number of sequenced proteins by 1980s was the development of automated sequencers followed by advances in mass spectrometry for structure identities. Structural knowledge is vital for complete understanding of life at the molecular level. An understanding of protein structure can lead to derivation of functions and mechanisms of their action. Bioinformatics is a novel approach in recent investigations on sequence analysis and structure prediction of proteins. With the advent of bioinformatics, it has been made possible to understand the relationship between amino acid sequence and three-dimensional structure in proteins. The central challenge of bioinformatics is the rationalization of the mass of sequence information not only to derive efficient means of storage and retrieval of sequence data, but also to design more analysis tools. Thus, there is a continual need to convert sequence information into biochemical and biophysical knowledge; to decipher the structural, functional and evolutionary clues encoded in the language of biological sequences (Attwood & Parry-Smith, 2003). Protein sequence information is stored in databases made available in the public domain to access, analyse and retrieve sequence and structural data. In general, protein databases may be classified as Primary and Secondary databases, composite protein pattern databases and structure classification databases. Primary and secondary databases address different aspects of protein analysis, because they store different levels of protein information. Primary databases are the central repositories of protein sequences, while secondary databases are based on the analysis of sequences of the primary ones. Composite protein pattern databases have emerged with a view to create a unified database of protein families. Protein structure classification databases have been established based on the structural similarities and common evolutionary origins of proteins. A number of tools are also available for protein structure visualization and protein identification and characterization. Thus bioinformatics tools for protein analysis provide a wealth of information related to sequences and structures of proteins.

Use of computational tools is an essential kit for the biologist in this rapid pace of information technology. Eventually, tools and techniques for protein sequence analysis and further, the structure prediction, has become an integral study for protein biochemists. Random identification of protein structures based only on homology of proteins is by and large an ambiguous approach. Hence, a systematic analysis of the protein under study from its sequence annotations to its three-dimensional structure alignment is a feasible approach for the investigations of protein structure aided by computational networking and repositories available in the public domain. Thus, sequence data can be transformed to structural data by a line of database analyses. The identification of protein structures can be organized as a flow of information from protein characterization, primary structure analysis and prediction by database search; sequence alignment; secondary structure prediction; motifs, profiles, patterns and fingerprint search; modeling; fold structure analysis and prediction; protein structure visualization and analysis of structure classification databases to deposition of protein structures in the public domain. An identity of sequence similarity of query sequences with that of database sequences indicating homology derives the phylogenetic maps of the protein under consideration and reveals information on conserved patterns thereby predicting repeat folds among the proteins that have arisen from divergence or of convergence. Pattern recognition methods convey information on the characteristics of unique features of the protein as well as the identification of similar traits in other proteins.

However, it is noteworthy that identifying patterns and functions of proteins are still far from being perfect which are likely to result in false interpretations and assumptions. Hence, it is the expertise and the reasoning of the biologist to interpret protein and/or any sequence information in the light of physical and chemicals methods to determine structure predictions. The study of bioinformatics is an interdisciplinary approach which requires the skill sets of biologists, mathematicians, information analysts and software developers to design and develop computational methods for analysis of biological data. This is presumably the index of milestones in bioinformatics for a fruitful journey in the identification of protein structure. Hence, it can be correlated that bioinformatics is the hand tool in every biology laboratory for thorough investigations of proteins and their complements in establishing evolutionary hierarchy and in the identification of protein malfunctions by linking protein structure to its functions in health and disease, thereby opening possible avenues for genetic manipulations and undertake prophylactic measures.

2. Protein structure-an overview

Protein architecture is the fundamental basis of the living systems that coordinates the functional properties of cells to sustain life. Every metabolic action is dependent on a set (s) of proteins that function as chaperones, enzymes, cofactors, structural proteins etc. Hence, an understanding of protein structure is vital for implications in physiological and therapeutic investigations. Lesk (2001) and Whitford (2005) have provided much of the understanding on the structural aspects of proteins. Generally, proteins are made up of small units known as amino acids which form a polypeptide chain through formation of peptide bonds. Thus, amino acids are the building blocks of all proteins which are characteristic for each type of the protein imparting specific functional attributes. There are 20 amino acids in nature that are of L-configuration that make up all kinds of proteins and are classified as aliphatic, aromatic, acidic, basic, hydroxylic, sulphur-containing and amidic amino acids. The discussion on the structure and chemical properties of amino acids is out of scope of this chapter and detailed information can be referred in most books covering protein structure. At the outset, we describe here the Primary, Secondary, Tertiary and Quaternary structures of a protein to enable keen insights of the structure prediction of proteins through bioinformatics. We also provide here the basic concepts of peptide bond and the Ramachandran plot that influence protein structure and conformation.

2.1 Primary structure

The primary structure of a protein resides in the linear order of the amino acid sequence along the polypeptide chain. Amino acids have been named in a three-letter code and in recent years, by a single letter code (Table 1) which is in current practice.

Amino acids	Three-letter code	Single letter code	Amino acids	Three-letter code	Single letter code
Alanine	Ala	A	Leucine	Leu	L
Arginine	Arg	R	Lysine	Lys	K
Asparagine	Asn	N	Methionine	Met	M
Aspartic acid	Asp	D	Phenylalanine	Phe	F
Cysteine	Cys	C	Proline	Pro	P
Glutamine	Gln	Q	Serine	Ser	S
Glutamic acid	Glu	E	Threonine	Thr	T
Glycine	Gly	G	Tryptophan	Trp	W
Histidine	His	H	Tyrosine	Tyr	Y
Isoleucine	Ile	I	Valine	Val	V

Table 1. Notations of amino acids in three-letter and single letter codes.

The amino acids that form the sequence are termed residues to denote the composition of a polypeptide. The primary sequence of a protein can therefore be visualized as a single letter code running from left to right with the left end constituting the N-terminal (amino group) of the first amino acid residue and the right end constituting the C-terminal (carboxylic acid group) of the last amino acid residue.

A particular amino acid residue of the amino acid sequence can therefore be identified by its position in the numerical sequence order. For example, a lysine residue can be identified as K6 when it appears in it 6th position or a glycine residue as G3 when it appears in its 3rd position and so on. The order of amino acid sequences is characteristic of a particular protein and of species and among protein families forming a conserved set of sequence in a region of the polypeptide(s). This sequential order determines the fold of a protein in achieving its native conformation and assigns the specific protein function. The primary sequence determination is therefore a significant criterion which defines the subsequent levels of the protein organization. An important aspect of the primary structure is that any mismatch of the sequence in a functional protein is often lethal to the cellular function carried out by the protein. This leads to several hereditary and metabolic defects such as in sickle cell anemia where the glutamic acid is replaced by valine in the 6th position of the β-chain of hemoglobin by a point mutation. The amino acid sequence of a protein is specified by the gene sequence by the process of transcription and translation.

2.2 Secondary structure
The secondary structure of a protein is the local conformation of the polypeptide chain or the spatial relationship of the amino acid residues which are placed close together in the primary sequence. This organizational level is found in globular proteins where three basic units of secondary structure are present, namely, the α-helix, β-strand and turns. Other secondary structures are based on these elements (Augen, 2004).

2.2.1 The α-helix
The right-handed α-helix is the most identifiable unit of secondary structure and the most common structural motif found in proteins with over 30% helix structure in globular proteins. In an α-helix, four or more consecutive amino acid residues adopt the same conformation resulting in a regular helical shape in the polypeptide backbone. This helix is stabilized by H-bonds between the main chain C=O group of one amino acid and the H-N group of the amino acid four residues further along the helix, forming a helix with 3.6 amino acid residues per helical turn resulting in a regular stable arrangement. The α-helix repeats itself every 0.54 nm along the helix axis i.e., the α-helix has a pitch of 0.54 nm. The radius of the helix is 0.23 nm with a translation distance per residue of 0.15 nm. The peptide planes are almost parallel with the helix axis and the dipoles within the helix are aligned. The α-helix arises from regular values adopted for φ (phi) and ψ (psi), the torsional or dihedral angles. The values of φ and ψ formed in the α-helix allow the backbone atoms to pack close together with few unfavorable contacts. This arrangement allows the H-bonding important for the stability of the helix structure. All the amino acids in the helix have negative φ and ψ angles, with ideal values of -57 and -47 respectively. It is important to note that proline does not form a helical structure due to the absence of an amide proton (NH) which is unable to form H-bond while the side chain covalently bonded to the N atom restricts backbone rotation.

2.2.2 The β strand
The second unit of protein secondary structure identified after the α-helix is the β strand which is an extended conformation when compared to the α-helix with 2 residues per turn and a translation distance of 0.34 nm leading to a pitch of nearly 0.7 nm in a regular β strand. A single β strand is not stable largely because of the limited number of local stabilizing interactions. When two or more β strands form additional H-bonding interactions, a stable sheet-like arrangement is created contributing to the overall stability of the β sheets. Adjacent strands can align in parallel or antiparallel arrangements and their orientations are established by the direction of the polypeptide chain from the N- to the C-terminal. Amino acid residues in the beta-conformation have negative φ and positive ψ angles with -139 and +135 angles respectively for parallel β sheets and -119 and +113 φ and ψ angles respectively for antiparallel β sheets. Polyamino acids in solution do not form β sheets and this hinders the study of their structures.

2.2.3 Turns
A turn is a secondary structural element where the polypeptide chain reverses its overall direction. It is a structural motif where the Cα atoms of two residues are separated by 1 to 5

peptide bonds and the torsional angles are not constant for all the residues in a turn. Many different conformations exist on the basis of the number of residues making up the turn and the dihedral angles associated with the central residues. Turns are classified according to the number of residues they contain namely, the α-turn, where the end residues are separated by 4 residues, β-turn, by 3 residues, γ-turn, by 2 residues, δ-turn, by one residue and π-turn, by a factor of 5 residues. A β-hairpin turn occurs between two H-bonded antiparallel beta strands in which the direction of the protein backbone reverses.

2.2.4 Loop
A loop occurs between 6 and 16 residues to form a compact globular shape of the protein which contain polar residues and hence, predominantly occur at the protein surface which contribute to the formation of active sites for ligand binding or catalytic activity. The loops connect the secondary structure elements of the polypeptide chain. Loop structures that are random are less stable and referred as random coils.

2.2.5 Coiled coil
A coiled coil is a structural motif in proteins in which 2-7 alpha helices are coiled together to form a repeated pattern of hypdrophobic and charged amino acid residues referred as heptad repeat. The tight packing in a coiled coil interface is due to van der Waal interactions between side chain groups. The coiled coil element is responsible for the amphipathic structures.

2.3 Tertiary structure
Tertiary structure is the global three-dimensional folding that results from interactions between elements of secondary structure. Tertiary structure of a protein therefore represents the folded conformation of a polypeptide chain in three-dimensional space, i.e., the spatial arrangement of amino acid residues widely separated in its primary structure. Interaction between the side chain groups is the predominant driver of the fold of the protein chain. These interactions which stabilize the tertiary structure arise from the formation of disulfide bridges, hydrophobic effects, charge-charge interactions, H-bonding and van der Waal interactions.
Disulfide bridges form between thiol (-SH) groups of two nearby cysteine residues. With reference to hydrophilic/hydrophobic interactions, water soluble proteins fold to expose hydrophilic side chains on the outer surface retaining the hydrophobic residues in the interior of the protein. Charge-charge interactions occur when a charged residue is paired with a neutralizing residue of opposite charge forming a salt bridge. H-bonding contributes to the overall stability of the tertiary structure or the folded state by stabilization of the secondary structure involving α-helices and parallel or antiparallel β sheets and of side chain groups of Tyr (Y), Thr (T), Ser (S), Gln (Q) and Asn (N). Van der Waal interactions are important in protein folding occurring between adjacent, uncharged and non-bonded atoms. A variety of post-translational modifications also contributes to the protein conformation such as conversion of proline to hydroxyproline that influences the tertiary structure of collagen molecule while glycosylation, carboxylation and methylation have little or no effects but which alter the chemical properties of the protein. Another important aspect in a protein fold is the activation of inactive proteins by small molecules such as cofactors, which are essential for native conformation formation.

The tertiary structure may be organized around more than one structural unit, known as domains which are folded sections of the protein representing structurally distinct units and the same interactions govern its stability and folding. Most domain structures exhibit specific functions independent of the rest of the protein architecture. Domain regions may be α-helices or β strands or mixed elements of both α-helices and β strands. Motifs are smaller structures, usually composed of few secondary elements that recur in many proteins and are rarely structurally independent. This feature or structural significance is important when considering the prediction of folded structure of an individual motif in context of the rest of a protein unlikely of the domain structure.

2.4 Quaternary structure

Many proteins involved in structural or metabolic or enzymatic functions are oligomeric proteins because they consist of more than a single polypeptide chains referred as subunits. The quaternary conformation of a protein arises from the interactions similar to tertiary structures, but is a result of interactions between the subunits which may be identical or nonidentical. Therefore, the quaternary structure refers to the noncovalent, stable association of the multiple subunits. A classic example of a protein that exhibits quaternary conformation is hemoglobin which consists of 4 polypeptide chains or subunits. The quaternary conformation of a protein allows the formation of catalytic or binding sites at the interface between subunits, which is not possible for monomeric proteins. Ligand or substrate binding causes a conformational change affecting the protein assembly for regulation of its biological activity such as the allosteric regulation in enzymes.

Thereby, the four conformations of a protein molecule define its architectural arrangement in a three-dimensional model which contribute to the functional attributes of the protein. This is represented in Figure 1 which is a common theme for most globular proteins.

2.5 The peptide bond

Amino acids are joined to each residue along the sequence by a linkage of the amino group of one residue with the carboxyl group of the next residue, known as the peptide bond (Figure 2).

The physical characteristics of the peptide bond impart the specific folding properties of the protein and this folding pattern of the polypeptide chain is described in terms of the angles of internal rotation around the bonds in the main chain. The N-Cα and Cα-C are single bonds and the internal rotations around these bonds are not restricted by the electronic structure of the bond, but, only by possible steric collisions in the conformations produced. An important characteristic of the peptide bond is the rigidity of the bond caused by its relatively short length, which imparts a partial double bond character. Hence, peptide bonds are characterized by a lack of rotational freedom. The double bond character of the peptide bond (Table 2) was first recognized by Linus Pauling who suggested that the peptide bond is rigid planar (Table 3) and hence exists as cis or trans isomer, with the trans isomer stable.

The entire conformation of the protein is described by these angles of internal rotation. Peptide bonds are invariably fixed at $\omega = 180°$. The φ and ψ angles are limited by steric hindrance between amino acid side chains which reduce the number of allowed conformations for a polypeptide chain. The rigidity of the peptide bond limits the number of arrangements that could fit without distorting the bonds. Without this constraint, the peptide would be free to adopt many numbers of structures and no single consistent pattern could exist. Therefore, by reducing the degrees of freedom, a well defined set of states of the protein could emerge. This is particularly significant because the proteins should indeed have a defined conformation to accomplish its physiological functions.

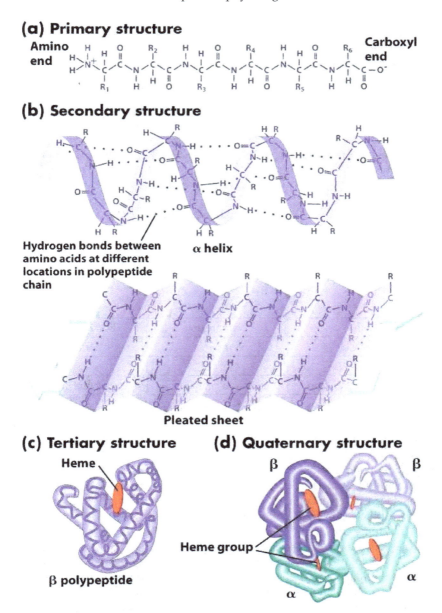

Fig. 1. Hierarchy levels of protein structure. The figure represents the different levels of hemoglobin structure.

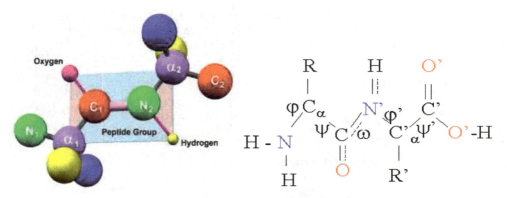

Fig. 2. The peptide bond structure. A. Ball and stick model, B. Torsional angles of the peptide structure

Bond nature	Length
C-N	1.47 Å
C=N	1.27 Å
C=O to NH	1.33 Å

Table 2. Bond character of the peptide bond.

Bond	Rotation	Torsional angle
NH to Cα	Free	Phi φ
Cα to C=O	Free	Psi ψ
C=O to NH (peptide bond)	Rigid planar	Omega ω

Table 3. Conformational angles of folding of polypeptide chain.

2.6 The Ramachandran plot

The peptide bond is planar as a result of resonance and its bond angle, ω has a value of 0 or 180°. A peptide bond in the trans conformation (ω =180°) is favoured over the cis arrangement (ω =0°) by a factor of ~1000 because the preferential arrangement of non-bonded atoms lead to fewer repulsive interactions that otherwise decrease stability. In the cis peptide bond these non-bonded interactions increase due to the close proximity of side chains and Cα atoms with the preceeding residue and hence results in decreased stability relative to the trans state. Peptide bonds preceeding Proline are an exemption to this trend with a trans/cis ratio of ~4. The peptide bond is relatively rigid, but far greater motion is possible about the remaining backbone torsion angles. In the polypeptide backbone C-N-Cα-C defines the torsion angle φ whilst N-Cα-C-N defines ψ. In practice these angles are limited by unfavourable close contacts with neighbouring atoms and these steric constraints limit the conformational space that is sampled by the polypeptide chains. The allowed values for φ and ψ were first determined by G.N.Ramachandran using a 'Hard sphere model' for the atoms and these values are indicated on a two-dimensional plot of φ against ψ that is called a Ramachandran plot.

In the Ramachandran plot shown in Figure 3 the freely available conformational space is shaded in green. This represents ideal geometry and is exhibited by regular strands or helices. Analysis of crystal structures determined to a resolution of <2.5 Å showed that over 80 percent of all residues are found in this region of the Ramachandran plot. The yellow region indicates areas that although less favourable can be formed with small deviations from the ideal angular values for φ and ψ. The yellow and green regions include 95 percent of all residues within a protein. Finally, the purple coloured region, although much less favourable will account for 98 percent of all residues in proteins. All other regions are effectively disallowed with the minor exception of a small region representing left handed helical structure. In total only 30 percent of the total conformational space is available suggesting that the polypeptide chain itself imposes severe restrictions. One exception to this rule is Glycine. Glycine lacks a Cβ atom and with just two hydrogen atoms attached to the Cα centre, this residue is able to sample a far greater proportion of the space represented in the Ramachandran plot. For glycine, this leads to a symmetric appearance for the allowed regions. As expected residues with large side chains are more likely to exhibit unfavourable, non-bonded interactions that limit the possible values of φ and ψ. In the Ramachandran plot the allowed regions are smaller for residues with large side chains such as phenylalanine, tryptophan, isoleucine and leucine when compared with the allowed regions for alanine.

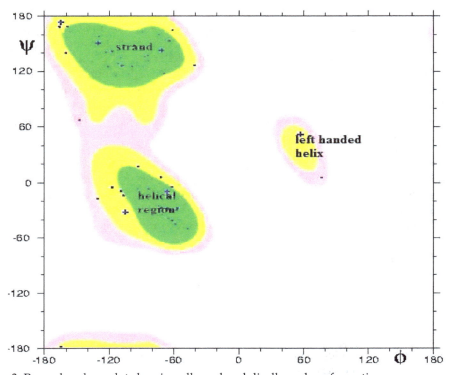

Fig. 3. Ramachandran plot showing allowed and disallowed conformations

3. The need for structural bioinformatics

Proteins are manifested in every aspect of biological activity/function. Many metabolic, cellular and structural events require the proper functioning of proteins in a cell. Any rupture of cellular function stems from the distortion or misfolding of proteins that prevents its normal function. Hence, protein science augments the advancements in genomic sciences to understand health and disease at the molecular level. Years back, an understanding of the structure of proteins, their interactions with other biomolecules, their roles within different biological systems have been made possible through molecular genetics and chemical methods and through biochemical pathways. This has taken years of intensive efforts and with the advent of modern techniques. The recent surge in bioinformatics has created a landmark in deciphering and decoding the gene and protein characteristics and functions. During the past decade, sequence information has been on a tremendous rise in contrast to the three-dimensional structural elucidation of proteins. This has resulted in the sequence/structure deficit of protein sequence and structure information. This can be estimated by the number of sequences available in sequence databases in contrast to the number of structures available in structure databases. A search for a protein sequence would generate hundreds of thousands of sequences while it would generate a few possible structures in a structure repository such as the Protein Data Bank (PDB) for the same protein query. This has prompted several consortia of groups to identify and deposit new protein structures through bioinformatics from the largely available protein structure prediction tools in the WWW.

Structure prediction has fascinated protein biochemists and the pioneering work of Margaret Dayhoff has contributed much to the understanding of protein structure through computational methods. She had developed the one-letter code for protein naming to reduce the complexity of the three-letter naming in the development of sequence information, storage and retrieval. She initiated the collection of protein sequences in the Atlas of Protein Sequence and Structure, a book collecting all known protein sequences that she published in 1965 which led to the development of Protein Information Resource database of protein sequences. In general, structure prediction is an attempt to predict the relative position of every protein atom in three-dimensional space using only its sequence information. Structural bioinformatics of the protein structure is based on a hierarchy of tools and techniques that identify the different levels of protein architecture (Figure 4). Many web tools for protein structure prediction have arisen to simplify the tasks of biochemists and bioinformaticians as well. Figure 5 provides a bird's eye view of the sequential steps in the identification/prediction of the protein structure.

4. Protein databases

Protein sequence information has been effectively dealt in a concerted approach by establishing, maintaining and disseminating databases, providing user-friendly software tools and develop state-of-the-art analysis tools to interpret structural data. Databases are central, shareable resources made available in public domain and represent convenient and efficient means of storing vast amount of information. Depending on the nature of the different levels of information, databases are classified into different types for the end user. This section describes the various databases for each of the nature of protein information that range from primary, composite, secondary and pattern databases. The different databases address different aspects of protein information which enable the analyst to perform an effective structure prediction strategy (Mala & Takeuchi, 2008).

4.1 Primary protein databases
4.1.1 PIR
This is the Protein Information Resource developed as a Protein sequence database at the National Biomedical Research Foundation (NBRF) in the early 1960s and collaboratively by PIR-International since 1988. The consortia include the PIR at NBRF, JIPID the International Protein Information Database of Japan and MIPS the Martinsried Institute for Protein Sequences.

4.1.2 MIPS
The Martinsried Institute for Protein sequences collects and processes sequence data for PIR and can be accessed at its web server.

4.1.3 SWISS-PROT
This protein database was produced collaboratively by the Department of Medical Biochemistry at the University of Geneva and the EMBL (European Molecular Biology Laboratory). Since 1994, it moved to EMBL's UK outstation, the EBI (European Bioinformatics Institute) and in April 1998, it moved to Swiss Institute of Bioinformatics (SIB) and is maintained collaboratively by SIB and EBI/EMBL. It provides the description of the function of proteins, structure of its domains, post-translational modifications etc., is minimally redundant and is interlinked to many other resources.

4.1.4 TrEMBL
This database has been designed to allow rapid access to protein sequence data. TrEMBL refers to Translated EMBL and was created as a supplement to SWISS-PROT in 1996 to include translations of all coding sequences in EMBL.

4.1.5 NRL-3D
This database is a valuable resource produced by PIR from sequences extracted from the Brookhaven Protein Data Bank (PDB). The significance of this database is that it makes available the protein sequence information in the PDB for keyword interrogation and for similarity searches. It includes bibliographic references, MEDLINE cross-references, active site, secondary structure and binding site annotations.

4.2 Composite protein sequence databases
Composite databases have been created to simplify the sequence search for a protein query in a single compilation in context of the many different primary database searches, by merging a variety of different primary resources. These databases are non-redundant and render sequence searching much more efficient.

4.2.1 NRDB
Non-Redundant DataBase (NRDB) is the default database of the NCBI (National Center for Biotechnology Information) BLAST (Basic Local Alignment Search Tool) service and is a

composite of GenPept, PDB sequences, SWISS-PROT, SPupdate (weekly update of SWISS-PROT), PIR and GenPept update (daily updates of GenPept). It provides comprehensive up-to-date information and is non-identical rather than non-redundant, that is, it reiterates only identical sequence copies and hence results in artifacts.

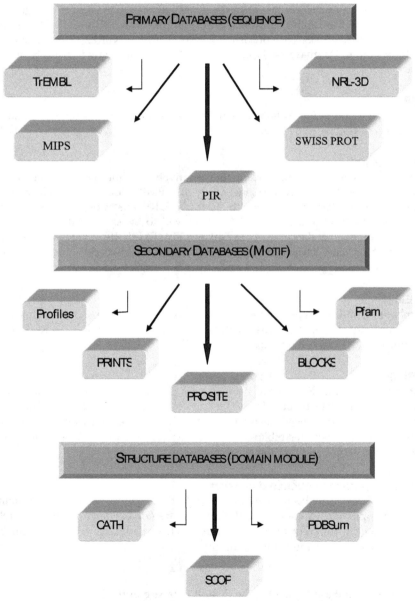

Fig. 4. Protein databases addressing different levels of protein structural information.

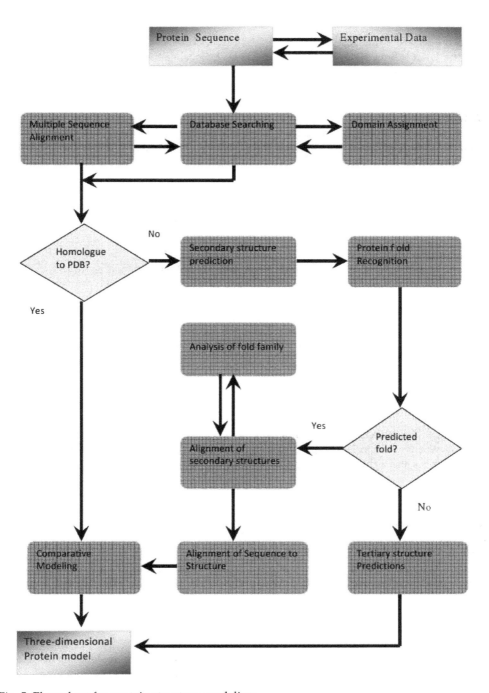

Fig. 5. Flow chart for protein structure modeling

4.2.2 OWL
This is a composite database of SWISS-PROT, PIR, GenBank and NRL-3D and is available from the UK EMBnet National Node and the UCL Specialist Node. It is a non-redundant database and is however not an updated resource but an efficient database for sequence comparisons.

4.2.3 MIPSX
This is a merged database produced at the Max-Planck Institute in Martinsried and reiterates unique copies of protein sequence search by removing identical sequences within or between them.

4.2.4 SWISS-PROT + TrEMBL
It is a combined resource of SWISS-PROT + TrEMBL at the EBI and is minimally redundant. It can be searched at the SRS sequence retrieval system on the EBI webserver.

4.3 Secondary databases
Secondary databases are a consequence of analyses of the sequences of the primary databases, mainly based from SWISS-PROT. Such databases augment the primary database searches, derived from multiple sequence information, by which an unknown query sequence can be searched against a library of patterns of conserved regions of sequence alignments which reflect some vital biological role, and based on these predefined characteristics of the patterns, the query protein can be assigned to a known family. However, secondary databases can never replace the primary sources but supplement the primary sequence search.

4.3.1 Prosite
It is the first secondary database and consists of entries describing the protein families, domains and functional sites as well as amino acid patterns, signatures, and profiles. This database was created in 1988 and is manually curated by a team of the Swiss Institute of Bioinformatics and tightly integrated into Swiss-Prot protein annotation.

4.3.2 Prints
This is a compendium of protein fingerprints. A fingerprint is a group of conserved motifs used to characterize a protein family by iterative scanning of a SWISS-PROT/TrEMBL composite. Usually the motifs do not overlap, but are separated along a sequence, though they may be contiguous in 3D-space. Fingerprints can encode protein folds and functionalities more flexibly and powerfully than can single motifs. PRINTS can be accessed by Accession number, PRINTS code, database code, text, sequence, title, number of motifs, author or query language.

4.3.3 Blocks
Blocks are multiply aligned ungapped segments corresponding to the most highly conserved regions of proteins. The blocks for the Blocks database are made automatically by looking for the most highly conserved regions in groups of proteins documented in InterPro. Results are reported in a multiple sequence alignment format without calibration and in the standard Block format for searching.

4.3.4 Profiles
In the motif-based approach of protein family characterization, it is probable that variable regions between conserved motifs also contain valuable sequence information. Profiles indicate where the insertions and deletions are allowed in the complete sequence alignment and provide a sensitive means of detecting distant sequence relationships.

4.3.5 Pfam
The Pfam database contains information about protein domains and families. For each entry a protein sequence alignment and a hidden Markov model is stored. These hidden Markov models can be used to search sequence databases. For each family in Pfam it is possible to look at multiple alignments, view protein domain architectures, examine species distribution, follow links to other databases and view known protein structures.

4.3.6 Identify
This resource is derived from BLOCKS and PRINTS and its search software eMOTIF is based on the generation of consensus expressions from conserved regions of sequence alignments. It can be accessed via the protein function webserver from the Department of Biochemistry at Stanford University.

4.4 Structure classification databases
Many proteins share structural similarities, reflecting common evolutionary origins. It can therefore be presumed that when the functions of proteins are conserved, the structural elements of active site residues may also be conserved giving rise to different fold families. Thus structure classification databases have evolved to better understand sequence/structure relationships. Important protein structure classification schemes are the CATH (Class, Architecture, Topology, Homology), SCOP (Structural Classification of Proteins) databases which will be dealt in detail in Section 9 of this Chapter.

4.5 Weblinks for protein databases
PIR   http://pir.georgetown.edu/
SWISS-PROT   http://expasy.org/sprot/
PROSITE   http://expasy.org/prosite/
PRINTS   http://www.bioinf.manchester.ac.uk/dbbrowser/PRINTS/index.php
Pfam   http://pfam.sanger.ac.uk/
SCOP   http://scop.mrc-lmb.cam.ac.uk/scop/
CATH   http://www.cathdb.info/

5. Sequence alignment
Two or more sequences share sequence similarities when they are homologous and share an ancestral sequence due to molecular evolution. Homology arises when the sequences share a common ancestor although similarity does not necessarily reflect homology below a certain threshold. When sequences exhibit similarities, it is likely that they will exhibit similarity of structures as well as biological functions, which enable to make predictions. This is the ultimate aim of sequence databases which requires the use of search tools that searches the sequences in the entire database against the new sequence or the query that has

been input by the user. Multiple alignments of protein sequences help to demonstrate homology which would otherwise have been considered non-significant in a pairwise alignment. In contrast to the homology of sequences over the entire length, it is also desirable to restrict homology to a limited region of the sequences. This is achieved by using a local alignment search tool, more commonly, the BLAST tool at NCBI. Multiple alignment tools are provided by EBI known as ClustalW program, most widely used with default and editable options in performing a multiple alignment (Figure 6).

BLAST is a heuristic method to find the highest scoring locally optimal alignments between a query sequence and a database sequence. It has been designed for fast database searching with minimal sacrifice of sensitivity and finds patches of local similarity, rather than a global fit. This tool works on statistics of ungapped sequence alignments and uses a substitution matrix in all phases of sequence searches. The use of filters reduces the artifacts in the databases. The BLAST algorithm works in a three-step process- the preprocessing of the query, generation of hits and extension of the hits. For a protein query, one can perform the standard BLASTP (a protein query vs. a protein database), TBLASTN (a protein query vs. six-frame translation of nucleotide sequences in the database), pairwise BLAST (between the first protein query sequence vs. the second protein sequence), PHI-BLAST (Pattern hit initiated BLAST which locates other protein sequences that contain both the regular expression pattern and the homologous sequences to a protein query) and the PSI-BLAST (Position specific iterated BLAST for finding protein families to determine domain identification and fold assignment).

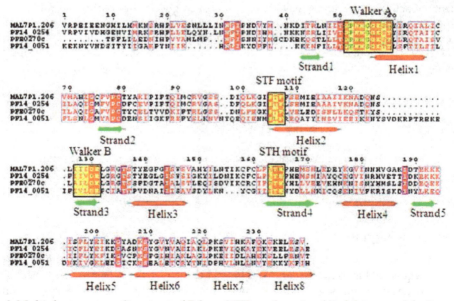

Fig. 6. Multiple sequence alignment of P-loop NTPase domain of *P. falciparum* MutS proteins. Conserved residues are in solid red and characteristic motifs are boxed in black and shaded in yellow. The corresponding secondary structure is shown below the alignment. Red cylinder represents helices and green arrows represent β-strands.

Comparing each and every sequence to every other sequence is an impractical means to obtain sequence similarity data. Often it is desirable to compare sequence sets of a given protein among its species and this is accomplished by a multiple sequent alignment by comparing all similar sequences in a single compilation, where, the sequences are aligned on top of each other, so that a co-ordinate system is set up. Each row corresponds to the sequence of a protein and each column is the same position in each sequence. Gaps are shown by dash '-' or dot '.' character. CLUSTALW is a standard program and W represents a specific version of the program. This program computes the pairwise alignments for all against for all sequences and the similarities are stored in a matrix. It converts the sequence similarity matrix values to distant measures, reflecting evolutionary distance between each pair of the sequences. It constructs a tree using neighbour-joining clustering algorithm and it progressively aligns the sequences/alignments together into each branch point of the tree. Clustal accepts alignments in several formats as: EMBL/SWISS-PROT, NBRF/PIR, GCG/MSF and its own format. There are 50 residues per line with one blank after 10 residues.

6. Protein data bank (PDB)

The Protein Data Bank (PDB) is the collection of structures and structural data of proteins, nucleic acids and other biological macromolecules. It was established in 1971 as a repository for the 3-D structural data at the Brookhaven National Laboratory, New York, and is available freely in the public domain. It is a key resource in the area of structural biology and structural genomics. PDB structures are deposited by researchers worldwide derived typically from X-ray crystallography, NMR spectroscopy, cryoelectron microscopy and theoretical modeling. PDB therefore serves as a platform to collect, organize and distribute structural information. Since 1998, PDB is an International Organization, managed by the Research Collaboratory for Structural Bioinformatics (RCSB) which facilitates the use and analysis of structural data in biological research. The PDB is overseen by an organization called the Worldwide Protein Data Bank, wwPDB. The founding members are PDBe (Europe), RCSB (USA) and PDBj (Japan). The BMRB (Biological Magnetic Resonance DataBank) joined in 2006. Each of the four members can act as deposition, data processing and distribution centres for PDB data. The data processing refers to the fact that wwPDB staff review and annotate each submitted entry. The data are then automatically checked for plausibility. The PDB website and its 'mirrors' permit retrieval of entries in computer-readable form (Kothekar, 2004).

6.1 PDB search
The PDB can be accessed at its homepage in the WWW (http://www rcsb.org/pdb/home/home.do) and several ways are available for search analysis using PDB identification code (PDB ID), searching the text found in PDB files (SearchLite), searching against specific fields of information such as deposition date or author (SearchFields), by searching the status of an entry on hold or released (StatusSearch) and by iterating on a previous search.

6.2 PDB structure
The PDB archive contains atomic coordinates, bibliographic citations, primary and secondary structure information, crystallographic structure factors and NMR experimental data. There are various options to view, download and search for structural neighbours. A

set of coordinates deposited with the PDB is subjected to a set of standard stereochemical checks and translated into a standard entry format. Each PDB entry is assigned an identifier with the first character denoting its version number.

7. Structure prediction methods

Structure prediction is an important aspect in modern biology which helps in the understanding of the functions and mechanisms of the protein macromolecule in medicine, pharmacology and biotechnology. In view of the complexity of the elucidation of protein structure by experimental means, it is now possible to use bioinformatics approaches for predictions of the protein structure. A number of software programs are available for structure predictions and the reasoning of the biologist to assess the suitability of the tools for the nature of the protein whose structure is to be determined is critical. The present methods for protein structure prediction include homology or comparative modeling, fold recognition or threading and *ab initio* or the *de novo* structure predictions for the appropriate proteins (Westhead et al., 2003). The basic approaches of these methods are discussed.

7.1 Homology or comparative modeling

This method is based on the consideration that sequences that are homologous by at least 25% over an alignment of 80 residues adopt the same structure while sequences falling below a 20% sequence identity can have very different structure. An important consideration is that tertiary structures of proteins are more conserved than their amino acid sequences. This is especially significant if a protein is similar but has been diverged; it could still possess the same overall structure. If a sequence of unknown structure (the target or query) can be aligned with one or more sequences of known structure (the template) that maps residues in the query sequence to residues in the template sequence, then, it produces a structural model of the target. Thus, homology modeling of a protein refers to constructing an atomic-resolution model of the target protein from its amino acid sequence and an experimental three-dimensional structure of the template. Homology models can be useful to derive qualitative conclusions about the biochemistry of the query sequence, about why certain residues are conserved. The spatial arrangement of conserved residues may suggest whether a particular residue is conserved to stabilize the folding, to participate in binding some small molecule, or to foster association with another protein or nucleic acid. Homology modeling can produce high-quality structural models when the target and template are closely related. The homology modeling procedure can be broken down into four sequential steps: template selection, target-template alignment, model construction, and model assessment. Figure 7 describes the sequence for homology modeling of a query protein.

The first critical step is to locate possible template structures using standard sequence similarity search methods such as BLAST for which the structures are experimentally known by experimental methods such as by X-ray crystallography or NMR spectroscopy and is available in the database. One of the limitations of homology modeling is the lack of a template structure for most proteins which is hoped to be available in the next 10-15 years with the advancements in structural genomics. When the template structure has been obtained, it is now essential to align the sequences with the target sequences by using a multiple alignment tool. When the target and template sequences closely match with high percentage identities, then, a good model is generated. The alignment should be generally checked for conserved key structural and functional residues to prevent obvious alignment errors when there is a high percentage identity. Given a template and an alignment, the information contained therein must be used to generate a three-dimensional structural model of the target, represented as a set of Cartesian coordinates for each atom in the protein. Three major classes of model generation methods have been proposed-fragment assembly, segment matching and satisfaction of spatial restraints. Regions of the target sequence that are not aligned to a template are modeled by loop modeling. The coordinates of unmatched sections determined by loop modeling programs are generally much less accurate particularly if the loop is longer than 10 residues. Homology models without reference to the true target structure are assessed by statistical potentials or physics-based energy calculations which produce an estimate of the energy for the model being assessed. The assessment of homology models' accuracy when the experimental structure is known is direct, using the root-mean-square deviation (RMSD) metric to measure the mean distance between the corresponding atoms in the two structures after they have been superimposed.

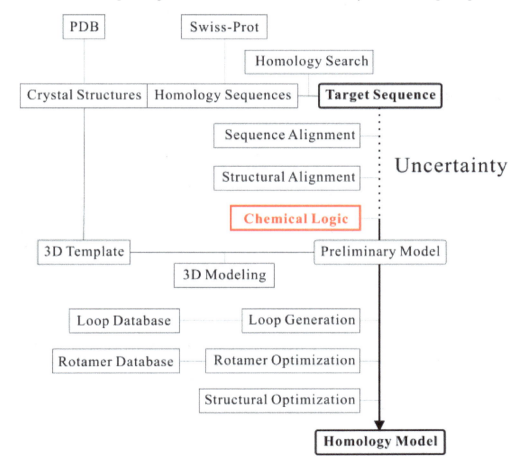

Fig. 7. Flow chart to derive protein structure by homology modeling

A number of free and commercial softwares are available in the WWW. SWISS-MODEL is a fully automated protein structure homology modeling server accessible via the ExPASy webserver or the SWISS-PDBVIEWER. It searches for suitable templates, checks sequence identity with targets, generates models and calculates energy minimization. MODELLER is another program for homology modeling. An alignment of the sequence to be modeled is to be provided and it automatically calculates a model with known related structures by satisfaction of spatial restraints. Table 4 lists some bioinformatics tools used for Homology modeling.

Web tool	Method
CABS	Reduced modeling tool
MODELLER	Satisfaction of spatial restraints
ROSETTA	Rosetta homology modeling
SWISS-MODEL	Local similarity / fragment assembly
TIP-STRUCTFAST	Automated comparative modeling
WHATIF	Position specific rotamers

Table 4. Homology modeling tools

7.2 Fold recognition or threading

The basic concept of threading was a result of the observation that a large percentage of proteins adopt one of a limited number folds; 10 different folds account for 50% of the known structural similarities between protein superfamilies. Rather than finding the correct structure for a given protein for all possible conformations, the correct structure is likely to have already been observed and stored in a database. In cases where the target protein shares significant sequence similarity to a protein of known 3-D structure, the fold recognition is made simple just by sequence comparison to identify the correct fold. The method of threading is thus used to detect structural similarities that are not accompanied by sequence similarity. Therefore, when a protein displays less than 25% sequence similarity to that of a template, the threading method can be used to predict its structure. This is unlike the homology modeling where sequence similarity is sufficient to guarantee similarity in structure. It is also evident that structures are conserved than sequences during evolutionary processes. Fold recognition method detects such distant relationships by searching through a library of known protein structures known as the fold library. Threading works by using statistical knowledge of the relationship between the structures deposited in the PDB and the sequence of the target protein. The prediction is made by "threading" or aligning each amino acid in the target sequence to a position in the template structure, and evaluating how well the target fits the template. After the best-fit template is selected, the structural model of the sequence is built based on the alignment with the chosen template. The flowchart of threading follows from the selection of protein structures from protein structure databases such as PDB and SCOP by eliminating structures with high sequence similarities, designing a good scoring function to measure the fitness between target sequences and templates based on the knowledge of the known relationships between the structures and the sequences, aligning the target sequence with each of the structure templates by optimizing the designed scoring function, selecting the threading alignment that is statistically most probable as the threading prediction and constructing a structure model for the target by placing the backbone atoms of the target sequence at their aligned backbone positions of the selected structural template. Fold recognition methods can be broadly divided into those that derive a 1-D profile for each structure in the fold library and align the target sequence to these profiles, and those that consider the full 3-D structure of the protein template. Fold recognition methods are widely used and effective because there are a strictly limited number of different protein folds in nature, mostly as a result of evolution and also due to constraints imposed by the basic physics and chemistry of polypeptide chains, which authenticate the derived protein structure by this method. Homology modeling and threading are both template-based methods but the protein templates that they target are very much different. However, this method suffers from its limitations such as the weak fold recognition and domain problem in proteins with multiple domains. Table 5 lists a few of the bioinformatics tools used in Threading.

Web tool	Method
PSI-BLAST	Iterated sequence alignment for fold identification
3D-PSSM	3D-1D sequence profiling
SUPERFAMILY	Hidden Markov model
GenTHREADER	Sequence profile and predicted secondary structure
LOOPP	Multiple methods

Table 5. Tools for Threading method

7.3 *Ab initio* or *De novo* structure prediction

Proteins fold to attain a state of minimum thermodynamic free energy as in all physicochemical systems. This is exploited to predict the structure conformation of the protein by *ab initio* methods. Thus, this method does not require a template structure but attempts to predict tertiary structure from the sequences that govern protein folding. Therefore it uses the principles of theoretical calculations in statistical thermodynamics and quantum mechanics. The different *ab initio* methods are Molecular dynamics simulations, Monte Carlo simulations, Genetic algorithm simulations and lattice models. However, this method is not in practice when compared to homology modeling or fold recognition due to its complexity in its approach.

Table 6 indicates some of the *ab initio* structure prediction methods.

Web tool	Method
ROSETTA	Rosetta homology modeling and *ab initio* fragment assembly
Rosetta@ Home	Distributed-computing implementation of Rosetta algorithm
CABS	Reduced modeling tool

Table 6. *Ab initio* programs

7.4 Strategies in protein structure prediction

A set of guidelines can be followed to devise a protein structure prediction strategy. The first step in the structure prediction of the protein can be to identify the features that the protein can possess that can be examined by sequence alignment. The presence of coiled coils could be tested. A prior analysis of the target sequence with Interpro can reveal an overall domain structure. Comparative model is more suited in terms of accuracy; although it is possible only for a minority of the proteins. Fold recognition methods detect evolutionary relationships inclusive of the consequence of divergence, however with lower accuracies.

8. Secondary structure prediction

The secondary structure of a protein refers to a consecutive fragment in its sequence that corresponds to a local region showing distinct geometrical features. These structural elements form during the early stages of the folding process. Knowledge of protein secondary structural regions along the protein sequence is a prerequisite to model the folding process or its kinetics. The ability to predict the secondary structure is a critical aspect in the structure prediction of a protein. Therefore, it is possible to recognize the three-dimensional topology by comparing the successfully predicted secondary structural elements of a query protein with the database of known topologies. Recently, it has been reported that helices and strand structures are maintained by evolution and the formation of regular secondary structure is an intrinsic feature of random amino acid sequences (Schaefer et al., 2010). Many methods are based on secondary structure propensity which reflects the preference of a residue for a particular secondary structure. Early methods were the Chou-Fasman method and the GOR method, while, predictions from multiple-aligned sequences are the Neural network methods which are based on statistical analysis. Other methods include Machine learning methods and Lim's and Cohen's methods.

8.1 Chou-Fasman method

It is a statistical prediction method based on calculation of statistical propensities of each residue forming either α-helix or β-strand. These propensities are used to classify the residues into six classes depending on their likelihood of forming an α-helix, and six classes depending on their likelihood of forming a β-strand. The class designations are used to find areas of probable α-helix and β-strands in the protein sequences to be predicted. The probable areas are then modified by a series of rules to produce the final prediction. This method is somewhat arbitrary and does not relate to chemical or physical theories. An improved version of this method for protein secondary structure prediction has been developed by Chen et al. (2006).

8.2 GOR method

The GOR (Garnier-Osguthorpe-Robson) method is based on statistical principles and is well-defined. It is based on the idea of treating the primary sequence and the sequence of secondary structure as two messages related by a translation process, which is then examined by using information theory. Structure prediction depends on measuring the amount of information the residues carry about their secondary structure and other residues secondary structure. Also theoretically complex, it is simple in practice.

8.3 PHD

This method uses a two-layered neural network method for sequence-to-structure prediction. The input of this network is a frame of 13 consecutive residues. Each residue is represented by the frequencies in the column of multiple sequence alignment which corresponds to that residue. The residues in the homologous proteins that correspond to the residue in the query protein are selected and frequencies of each type of residues are calculated and input to the network. This means each residue introduces 20 inputs to the neural network. Also, one more input is used for each residue in the frame for the cases that the frame extends over the N or C terminus of the protein. One final input is added for each residue called the conservation weight. This weight represents the quality of a multiple sequence alignment. So every residue is represented by 20+1+1=22 inputs, thus the sequence-to-structure network has 13x22 input modes. The output of this network is 3 weights, one for each of the helix, strand and loop states. The structure-to-structure prediction part of the algorithm is also implemented as a two-layered feed-forward network (Singh et al., 2008).

8.4 Machine learning methods

The first full-scale application of machine learning to secondary structure prediction occurred with the development of a learning algorithm PROMIS. Since then, more powerful machine learning methods known as inductive logic programming (ILP) have been developed. ILP method is specifically designed to learn structural relationships between objects and is more advantageous for secondary structure prediction, using the database of known structures.

8.5

Lim's method and Cohen's method are based on physicochemical properties to encode structural knowledge of proteins.

8.6

Multiple sequence alignments significantly improve secondary structure prediction and reveal patterns of conservation as a result of evolution. A residue with a high propensity for a particular secondary structure in one sequence may have occurred by chance, but if it is part of a conserved column, in which all residues have high propensity for that type of secondary structure, then it provides predictive evidence. Multiple alignments can also reveal subtle patterns of conservation. Like, for example, a large proportion of α-helices in globular proteins are amphipathic, containing hydrophobic and hydrophilic residues associated with periodic patterns of sequences. The appearance of such conserved patterns is therefore predictive of α-helical structure.

8.7

Secondary structure prediction tools are the Jpred which is a neural network assignment method, PREDATOR which is a knowledge-based databse comparison method and Predict protein which is a profile-based neural network.

9. Structural classification

Protein structure is more conserved than its sequences. Hence, there is a need for classification of protein structures for management of protein structures deposited in databases to reflect both structural and evolutionary relatedness. Protein classification is based on a hierarchy of levels which assign the proteins to family, superfamily and fold depicting clear evolutionary relationship, probable common evolutionary origin and major

structural similarity respectively. Methods of protein structure classification rely on the sequence comparison methods and the structure comparison methods.
CATH and SCOP are the major hierarchical structure classification databases available at: http://www.cathdb.info/ and http://scop.mrc-lmb.cam.ac.uk/scop/ in the www.

9.1 CATH
This database classifies proteins based on its Class, Architecture, Topology and Homology. Class is determined by secondary structure and packing within the folded protein. Three major classes are recognized: all alpha, all beta and alphabeta, while the fourth class is composed of proteins with low secondary structure content. Architecture represents the overall shape of the domain as a function of the orientations of individual secondary structures. This level is assigned using a simple description of the secondary structure arrangement. The Topology level groups proteins into fold families depending on both the overall shape and connectivity of secondary structures. The Homologous superfamily level groups together protein domains that share a common ancestor. Structures within this level are further clustered according to their level of sequence identity.

9.2 SCOP
This database represents the Structural Classification of Proteins, a valuable resource for comparing and classifying new structures. It is designed to provide a comprehensive description of the structural and evolutionary relationships between all proteins whose structure is known, which includes all entries in the PDB. The database is available as a set of tightly linked hypertext documents for accessibility. This classification has been constructed manually by visual inspection and comparison of structures.

10. Structure visualization

Structure visualization enables identification and manipulation of structural features in the three-dimensional view of protein macromolecules. Several programs have been developed to view structural data. Rasmol is one of the most popular tools for protein structure visualization developed by Roger Sayle which reads molecular structure files from PDB. Chemscape Chime and Protein explorer work as plug-ins to allow structure visualization in the web browser. Cn3-D is a helper application that allows viewing of 3-D structures and sequence-structure or structure-structure alignments for NCBI database. Swiss-PdbViewer provides an interface to analyze several proteins at the same time, which can be superimposed in order to deduce structural alignments and compare their active sites. It is tightly linked to Swiss-Model. PDBsum is a database which provides a largely pictorial summary of the key information on each macromolecular structure from the PDB. Table 7 lists some of the databases used for protein structure visualization.

11. Web tools in protein structure prediction

There exists unlimited information in the WWW for determination of protein structure prediction due to developments of webservers to analyse and interpret structural data. Webservers are developed and maintained by the Organizations for free availability or on commercial purposes. With the recent revolutions in bioinformatics, new software tools have been designed to meet updated protein information. This section is therefore intended to describe some of the webservers for obtaining structural information.

Program	Function
RasMol	3-dimensional visualization
Cn3-D	3-dimensional visualization, linked to sequence alignments
Chime	3-dimensional visualization
TOPS	Visualization of protein folding topologies
DSSP	Finds secondary structure elements in an input structure
Surfnet	Visualization of protein surface
PROCHECK	Checks stereochemical quality of protein structures
PROMOTIF	Analyses protein structural motifs

Table 7. Protein visualization tools

11.1 ExPASy

The ExPASy (Expert Protein Analysis System) is a proteomics server of the Swiss Institute of Bioinformatics (SIB) which analyzes protein sequences and structures and functions in collaboration with the European Bioinformatics Institute. The ExPASy server is a repertoire of tools for the many different types of protein analysis. These tools can be accessed at: http://expasy.org/tools/ and retrieve information on protein identification and characterization using mass spectrometric data, primary structure analysis, pattern and profile searches, secondary structure prediction, tertiary sequence analysis and tertiary structure prediction as well as quaternary structure analysis, molecular modeling and visualization.

Table 8 lists some of the Protein identification and characterization programs (Mala & Takeuchi, 2008).

Table 9 lists the protein structure prediction programs (Rastogi et al., 2004).

Program	Function
AACompIdent	Identification of amino acid composition
TagIdent	Identification of proteins using mass spectrometric data
PeptIdent	Identification of proteins using peptide mass fingerprint data
MultiIdent	Identification of proteins using pI, MW, amino acid composition
Propsearch	Find putative protein family
PepSea	Identification of protein by peptide mapping or peptide sequencing
FindPept	Identification of peptides resulting from unspecific cleavage of proteins
TMAP; TMHMM	Prediction of transmembrane helices

Table 8. Tools in protein identification and characterization

Program	Function
ProtParam	Physico-chemical parameters of a protein sequence
HeliQuest	A webserver to screen sequences with specific alpha-helical properties
Rep	Searches a protein sequence for repeats
Paircoil	Prediction of coiled coil in proteins
PepDraw	Peptide primary structure drawing
Jpred	A consensus method for protein secondary structure prediction
PredictProtein	A webserver from Columbia University for secondary structure prediction
PSIpred	Various protein structure prediction methods
SWISS-MODEL	An automated knowledge-based protein modeling server
LOOPP	Sequence to sequence, sequence to structure, and structure to structure alignment
Rosetta	Prediction of protein structure from sequence
MakeMultimer	Reconstruction of multimeric molecules present in crystals
Swiss-PdbViewer	A program to display, analyse and superimpose protein 3D structures

Table 9. Tools in Protein structure prediction

11.2 Predict protein

PredictProtein is a webserver available at http://www.predictprotein.org/ and works on the profile-based neural network method. It integrates feature prediction for secondary structure, solvent accessibility, transmembrane helices, globular regions, coiled-coil regions, structural switch regions, B-values, disorder regions, intra-residue contacts, protein-protein and protein-DNA binding sites, sub-cellular localization, domain boundaries, beta-barrels, cysteine bonds, metal binding sites and disulphide bridges. PredictProtein caches the prediction for each new query sequence it processes for quick and easy retrieval. Currently the PredictProtein cache contains 4,136,476 annotated proteins.

11.3 Rasmol

The software developed by Roger Sayle displays a three-dimensional image of a structure in the standard structural database. The image can be rotated by using a mouse to produce different views and displayed in various formats such as wireframe, space filling, ball and stick and cartoon formats, which give clear displays of secondary structure elements. The user can choose between various colour schemes and even use customized colours. There are flexible ways of selecting parts of structures to enable highlighting with a different display format. Figure 8 shows the different displays of a protein structure viewed in Rasmol.

11.4 DOMO and PROF_PAT

A new database of aligned protein domains known as DOMO has been developed by Gracy & Argos (1998). DOMO can be accessed through the sequence retrieval system (SRS). A form-based query manager allows retrieval of familial domain alignments by identifiers, sequence accession numbers or keywords. The DOMO sequence analysis provides a simple tool for determining domain arrangements, evolutionary relationships and key amino acid residues in a query protein sequence. PROF_PAT 1.3 is an updated database of patterns to detect local similarities, containing patterns of more than 13,000 groups of related proteins in a format similar to that of PROSITE.

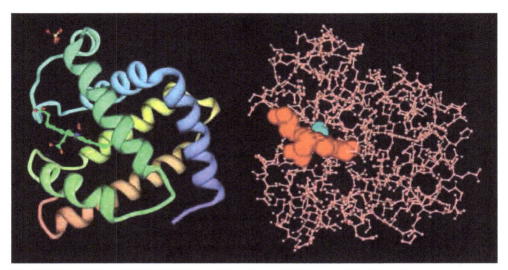

Fig. 8. Ribbon (A) and Ball stick (B) models of myoglobin viewed in Rasmol

12. CASP

CASP refers to Critical Assessment of protein Structure Prediction experimental methods to establish the current state of the art in protein structure prediction with identification of the progress made so far and highlight future efforts to be focused. CASP1 was initiated in 1994 and has been upgraded every two years. The recent method is CASP9 for the year 2010. CASP provides research groups with an opportunity to objectively test their structure prediction methods and delivers an independent assessment of the state of the art in protein structure modeling to the research community and software users. Prediction methods are assessed on the basis of the analysis of a large number of blind predictions of protein structure. The CASP results are published in special supplement issues of the scientific journal Proteins, all of which are accessible through the CASP website. The earlier version of CASP, CASP8 has been described byMoult et al. (2009) in Proteins.

13. Conclusion

Protein sequence information can be retrieved and analysed from databases that encompass much of the available sequence and structure data. On the other hand, it is of significant interest that a researcher be able to submit the sequence information for the protein investigated by him/her. Submission of sequences in any of the databases is transferred by FTP to the other databases for synchronized database management. The online submission tools provide a simple user interface and are maintained and curated on a daily basis. The vast sequence information available in the WWW requires potential search engines for data retrieval such as the Entrez from NCBI and SRS (sequence retrieval system) from EBI, which allow text-based searching of a number of linked databases. Thus, there is a continual need for sequence information and data retrieval in view of the sequence/structure deficit and also to provide links to the identification of protein biomarkers in health and disease which requires structural information. This Chapter therefore provides comprehensive information to the reader on the application-based insights of protein structure prediction using bioinformatics approaches.

14. Acknowledgement

The authors thank Dr. A.B.Mandal, Director, CLRI, Chennai, India, for his kind permission to publish this work. The financial assistance extended by the Council of Scientific and Industrial Research (CSIR), New Delhi, India, to Dr. J.Geraldine Sandana Mala is gratefully acknowledged. The financial support of Mr. Kikuji Takeuchi and Mr. Naomi Takeuchi of Takenen, Japan for publication of this Chapter is also sincerely acknowledged.

15. References

Attwood, T.K. & Parry-Smith, D.J. (2003) *Introduction to Bioinformatics*, Fourth Indian reprint, Pearson Education Ltd., ISBN 81-7808-507-0, India

Augen, J. (2004) *Bioinformatics in the Post-genomic era*, First edition, Pearson Education Inc., ISBN 0-321-17386-4, USA

Chen, H., Gu, F. & Huang, Z. (2006) Improved Chou-Fasman method for protein secondary structure prediction. *BMC Bioinformatics*, 7, pp.S14

Gracy, J. & Argos, P. (1998) DOMO: a new database of aligned protein domains. *TIBS*, 23, pp. 495-497

Kothekar, V. (2004) *Introduction to Bioinformatics*, First edition, Dhruv Publications, ISBN 81-8240-006-6, India

Lesk, A.M. (2001) *Introduction to Protein architecture*, First edition, Oxford University Press, ISBN 0-19-850474-8, USA

Mala, J.G.S. & Takeuchi, S. (2008) Understanding structural features of microbial lipases-an overview. *Analytical Chemistry Insights*, 3, pp. 9-19

Moult, J., Fidelis, K., Kryshtafovych, A., Rost, B. & Tramontano, A. (2009) Critical assessment of methods of protein structure prediction – Round VIII. *Proteins*, 77, pp.1-4

Rastogi, S.C., Mendiratta, N. & Rastogi, P. (2004) *Bioinformatics: Methods and Applications*, First edition, Prentice-Hall of India Pvt. Ltd., ISBN 81-203-2582-6, India

Schaefer, C., Schlessinger, A. & Rost, B. (2010) Protein secondary structure appears to be robust under *in silico* evolution while protein disorder appears not to be. *Bioinformatics*, 26, pp. 625-631

Singh, M., Sandhu, P.S. & Kaur, R.K. (2008) Protein secondary structure prediction. *World Academy of Science, Engineering and Technology*, 42, pp. 458-461

Westhead, D.R., Parish, J.H. & Twyman, R.M. (2003) *Instant Notes Bioinformatics*, First Indian edition, Viva Books Pvt.Ltd., ISBN 81-7649-419-4, India

Whitford, D. (2005) *Proteins structure and function*, John Wiley & Sons, ISBN 0-471-49893-9 HB, ISBN 0-471-49894-7PB, England

8

Systematic and Phylogenetic Analysis of the Ole e 1 Pollen Protein Family Members in Plants

José Carlos Jiménez-López,
María Isabel Rodríguez-García and Juan de Dios Alché
Department of Biochemistry, Cell and Molecular Biology of Plants
Estación Experimental del Zaidín, CSIC, Granada,
Spain

1. Introduction

Pollen allergens are specific substances able to cause IgE-mediated hypersensitivity (allergy) after contact with the immune system [D'Amato et al. 1998]. To date, about 50 plant species have been registered in the official allergen list of the International Union of Immunological Societies (IUIS) Allergen Nomenclature Subcommittee http://www.allergen.org as capable of inducing pollen allergy in atopic individuals [Mothes et al. 2004]. These plants are usually grouped as (1) trees (members of the orders: *Fagales, Pinales, Rosales, Arecales, Scrophulariales, Junglandales, Salicales,* and *Myrtales*), (2) grasses (members of the families: *Bambusioideae, Arundinoideae, Chloridoideae, Panicoideae,* and *Poideae*), and (3) weeds (components of families *Asteraceae, Chenopodiaceae* and *Urticaceae*) [Hauser et al. 2010].

Allergens are proteins with a broad range of molecular weights (~5 to 50 kDa), which exhibit different features of solubility and stability. More than 10 groups of pollen allergens have been reported. Among all groups of pollen allergens, Pollen Ole e I (Ole) domain-containing proteins are the major allergens, included like-members of the "pollen proteins of the Ole e 1 family" (Accession number: PF01190) within the Pfam protein families database [Finn et al. 2010].

Ole e 1 was the first allergen purified from *Olea europaea* L. [Lauzurica et al. 1998] and named as such according to the IUIS nomenclature [King et al. 1994]. This protein is considered the major olive pollen allergen on the basis of its high prevalence among atopic patients and the high proportion it represents within the total pollen protein content, in comparison with other olive pollen allergens. These include at present another 10 allergens already identified and classified like Ole e 2 to Ole e 11 [Rodríguez et al. 2002, Barral et al. 2004, Salamanca et al. 2010]. Ole e 1 consists of a single polypeptide chain of 145 amino acid residues with a MW of 18–22 kDa, displaying acidic pI and different forms of N-glycosylation [Villalba et al. 1990, Batanero et al. 1994]. Heterologous proteins with a relevant homology have been described in other members of the *Oleaceae* family, such a fraxinus, lilac, jasmine and privet. The polypeptides encoded by the *LAT52* gene from tomato and the *Zmc13* gene from maize pollens also exhibit a high similarity to Ole e 1 [Twell et al. 1989, Hanson et al. 1989]. These plant pollen proteins are structurally related but their biological function is not yet known; though they have been suggested to be

involved in important events of pollen physiology, such as hydration, germination and/or pollen tube growth, and other reproductive functions [Alché et al. 1999, 2004, Tang et al. 2000, Stratford et al. 2001].
Structurally, the Ole domain contains six conserved cysteines which may be involved in disulfide bonds, since no free sulfhydryl groups have been detected in the native protein [Villalba et al. 1993]. Olive Ole e 1 exhibits a high degree of microheterogeneity, mainly concentrated in the third of the molecule closer to the N- terminus. The Ole e I (Ole) domain defining the pollen proteins Ole e I family signature or consensus pattern sequences PS00925 [Sigrist et al. 2010], is characterized by the amino acid sequence [EQT]-G-x-V-Y-C-D-[TNP]-C-R, where "x" could be any residue.
There is a high diversity of proteins sharing the Ole domain among plant species. To date, eleven Ole domain-containing genes have been isolated and characterized from olive pollens [Rodríguez et al. 2002]. Ole-containing proteins include proline-rich proteins, proteins encoding extensin-like domains, phosphoglycerate mutase, tyrosine-rich hydroxyproline-rich glycoprotein, and hydroxyproline-rich glycoprotein. These Ole-containing proteins can exhibit: (1) the pollen Ole signature exclusively, e.g. the ALL1_OLEEU P19963 protein from *Olea europaea* L., (2) both the pollen Ole signature and the replication factor A protein 3 motive pattern (PF08661), e.g. the O49527 pollen-specific protein-like from *Arabidopsis thaliana* (842 residues), (3) both the pollen Ole domain and the phosphoglycerate mutase (PGAM) motif, e.g. the Q9SGZ6 protein from *Arabidopsis thaliana*., and finally (4) both the pollen Ole signature and the reverse transcriptase 2 (RVT2) motif, e.g. the A5AJL0 protein from *Vitis vinifera*.
Several efforts have been made to develop an understandable and reliable systematic classification of the diverse and increasing number of different allergen protein structures. As mentioned above, the classification system widely established for proteins that cause IgE-mediated atopic allergies in humans (allergens) was defined by Chapman et al. (2007). This system uses the first three letters of the genus; a space; the first letter of the species name; a space and an Arabic number. Despite this classification system, protein databases are full of allergen proteins lacking this systematic and comprehensive nomenclature. In other cases, many of the proteins described here have not been described as allergens, or their naming makes no reference to the Ole e 1 family that facilitates their identification. Otherwise, naming in databases is frequently given randomly, on the basis of chromosome location, addressing structural features and functional characterizations or simply using the name of the entire family. In this study, we used a combination of functional genomics and computational biology to name and classify the entire Ole e 1 family, as well as to characterize structurally and functionally the proteins of this superfamily. Our data indicate that the Ole e 1 protein family consists of at least 109 divergent families, which will likely expand as more genomic studies are undertaken, and fully sequenced plant genomes become available.

2. Material and methods

2.1 Database search for Ole e 1 family genes

Sequences of Ole e 1 and Ole e 1-like genes were retrieved from the US National Center for Biotechnology Information (NCBI, http://www.ncbi.nlm.nih.gov/), the Uniprot database (http://www.uniprot.org/), and the non-redundant expressed sequence tag (EST) databases using BLASTX, BLASTN and BLAST (low complexity filter, Blosum62 substitution matrix) [Altschul et al. 1997]. Searches were conducted using previously characterized *Olea europaea* L. *Ole e 1* (GenBank Accession number P19963), *Solanum lycopersicum LAT52* (GenBank Accession number P13447), *Zea mays Zmc13* (GenBank Accession number B6T1A9), *Arabidopsis thaliana* pollen-specific protein-like (GenBank Accession number O49527), *Arabidopsis thaliana* PGAM containing domain protein (GenBank Accession number Q9SGZ6), and *Vitis vinifera* RVT2 containing domain protein (GenBank Accession number A5AJL0). Full-length amino acid sequences for Ole e 1 proteins were compiled and aligned using ClustalW [Thompson et al. 1994]. Genetic distances between pairs of amino acid sequences were calculated with Bioedit V7.0.5.3 [Hall 1999]. Consensus protein sequences were derived from these original alignment, and further analyzed for the presence of putative functional motifs using the PROSITE database [Sigrist et al. 2010], of biologically meaningful motif descriptors derived from multiple alignments and the ScanProsite program [de Castro et al. 2006], from the Expert Protein Analysis System (ExPASy) proteomics server of the Swiss Institute of Bioinformatics [Gasteiger et al. 2003]. Finally, the consensus protein sequences were submitted to BLASTP analysis to identify homologous proteins from other plant species.

2.2 Revised/unified nomenclature

In order to provide a revised and unified nomenclature for Ole e 1-like gene superfamily, we developed a sequence-based similarity approach to classify all the retrieved sequences using a previously developed gene nomenclature model [Kotchoni et al. 2010]. For this new nomenclature, Ole e 1 protein sequences that are more than 40% identical to previously identified Ole e 1 sequences compose a family, and sequences more than 60% identical within a family, compose a gene subfamily. Protein sequences that are less than 40% identical would describe a new Ole e 1 gene family. Taking olive protein Ole e 1_57A9 (previous name Ole e 1, major olive pollen allergen) as an example for the revised nomenclature (Table 1), Ole e 1 indicates the root; the digits (57) indicates a family and the first letter (A) a subfamily, while the final number (9) identifies an individual gene within a subfamily. The revised nomenclature is therefore composed of an assigned gene symbol (Ole e 1) (abbreviated gene name) for the whole gene superfamily. The gene symbol must be (i) unique and representative of the gene superfamily; (ii) contain only Latin letters and/or Arabic numerals, (iii) not contain punctuation, and (iv) without any reference to species. These newly developed criteria have been applied to database curators to generate the unified Ole e 1 gene families/classes regardless of the source of the cloned gene(s).

2.3 Sequence alignments and phylogenetic analyses

The retrieved Ole e 1 protein families were used to generate a phylogenetic tree using ClustalW [Thompson et al. 1994]. The alignment was created using the Gonnet protein weight matrix, multiple alignment gap opening/extension penalties of 10/0.5 and pairwise gap opening/extension penalties of 10/0.1. These alignments were adjusted using Bioedit V7.0.5.3 [Hall 1999]. Portions of sequences that could not be reliably aligned were eliminated. Phylogenetic tree was generated by the neighbourjoining method (NJ), and the branches were tested with 1,000 bootstrap replicates. The three was visualized using Treedyn program [Chevenet et al. 2006].

2.4 Ole e 1 superfamily: Protein modeling and structural characterization

In order to study the structural and conformational variability between the Ole e 1 protein families, selected members of the Ole e 1 superfamily were modelled using SWISS-MODEL server, via the ExPASy web server [Gasteiger et al. 2003]. The initial modelled Ole e 1 structures were subjected to energy minimization with GROMOS96 force field energy [van Gunsteren et al. 1996] implemented in DeepView/Swiss-PDBViewer v3.7 [Guex and Peitsch 1997] to improve the van der Waals contacts and to correct the stereochemistry of the improved models. The quality of the models was assessed by checking the protein stereology with PROCHECK [Laskowski et al. 1993] and the protein energy with ANOLEA [Melo et al. 1997, 1998]. Ramachandran plot statistics for the models were calculated to show the number of protein residues in the favoured regions.

3. Results

3.1 The Ole e 1 protein families: Revised and unified nomenclature

In order to provide a revised/international consensus and unified nomenclature for the Ole e 1 gene superfamily, we first retrieved all the Ole e 1 and Ole e 1-like gene sequences using PS00925 as the major molecular consensus defining the entire superfamily of Ole e 1 proteins. We next verified all annotated plant Ole e 1 open reading frames (ORFs) using Ole e 1 sequence domains. A complementary and comparative study was developed by using Uniprot database to validate the molecular function and previous denomination of each Ole e 1 protein. Our searches resulted in the identification of 571 sequences encoding Ole e 1 and Ole e 1 like proteins from a wide variety of plant species, with the diagnostic motif PS00925 (Table 1). According to the established criteria (see Material and Methods), these sequences integrated 109 Ole e 1 gene families which have been attributed to different functional categories including extensins and extensin-like proteins, proline-rich proteins, hydroxyproline-rich glycoproteins, tyrosine-rich/hydroxyproline-rich glycoproteins, hydrolases, phosphoglycerate mutases, arabinogalactan proteins, etc. (Table 1).

Among the sequences retrieved, Ole e 1_48 is the most extensive family with 63 gene members encoding for different pollen-specific protein C13 homologues, followed by Ole e 1_57 family with 42 gene homologues encoding Ole e 1 (the olive major pollen allergen), Ole e 1_16 with 26 gene members encoding proline-rich proteins, and Ole e 1_52 with 22 members encoding LAT52 homologues (Table 1). The number of Ole e 1 genes greatly varied from one plant species to another. The genus *Oryza* included the highest number of Ole e 1 genes (143), followed by *Arabidopsis* with 95 genes (Table 1). At present, more than half of the catalogued Ole e 1 families encoded a single Ole e 1/Ole e 1-like gene, which was in most cases "uncharacterized" (Table 1).

The total number of genes in the Ole e 1 superfamily is expected to increase steadily with time, mainly due to the genomic sequencing of additional species like *Olea europaea* L. (http://www.gen-es.org/11_proyectos/PROYECTOS.CFM?pg=0106&n=1). Regardless of the plethora of Ole e 1 genes yet to be identified/characterized, their classification and relationship to the entire extended Ole e 1 gene superfamily will be easy owing to this nomenclature building block that catalogues newly identified/characterized Ole e 1 gene products only on the basis of sequence similarity to previously characterized Ole e 1 gene products.

Ole e 1 Family	Revised annotation	Previous annotation	GenBank Accession number	Source
1	Ole e 1_1A1	At4g1/215	Q8RXZ6	ARATH
1	Ole e 1_1A2	-	Q8L8V9	ARATH
1	Ole e 1_1A3	ARALYDRAFT_493155	D7MC15	ARALY
1	Ole e 1_1A4	40.t00006	Q2A9B5	BRAOL
1	Ole e 1_1A5	31.t00008	Q2A9F3	BRAOL
1	Ole e 1_1B1	ARALYDRAFT_483053	D7LEF7	ARALY
1	Ole e 1_1B2	At2g40113	Q5BFY6	ARATH
1	Ole e 1_1B3	-	Q9LE62	ARATH
1	Ole e 1_1B4	At5g47635	Q29PT1	ARATH
1	Ole e 1_1B5	ARALYDRAFT_330672	D7MPZ6	ARALY
2	Ole e 1_2A1	POPTRDRAFT_818026	B9HCD0	POPTR
2	Ole e 1_2A2	POPTRDRAFT_776772	B9H1F6	POPTR
2	Ole e 1_2B1	VIT_00005138001	D7SSK6	VITVI
2	Ole e 1_2C1	-	C6T3E9	SOYBN
2	Ole e 1_2D1	RCOM_0880370	B9SA60	RICCO
3	Ole e 1_3A1	OsJ_33016	B8BG44	ORYSI
3	Ole e 1_3A2	Os10g0206500	Q109X3	ORYSJ
3	Ole e 1_3B1	OSJNBa0014J14.3	Q7G7E7	ORYSJ
3	Ole e 1_3B2	OJ1084_F02.8	Q6RV11	ORYSJ
3	Ole e 1_3C1	OSJNBa0014J14.28	Q3S6U0	ORYSJ
3	Ole e 1_3D1	Os10g0209600	Q109X1	ORYSJ
3	Ole e 1_3D2	OsJ_33026	B8BG53	ORYSI
3	Ole e 1_3E1	SORBIDRAFT_01g013620	C5WR76	SORBI
3	Ole e 1_3F1	-	B4FE56	MAIZE
4	Ole e 1_4A1	SELMODRAFT_444624	D8SBK5	SELML
4	Ole e 1_4A2	SELMODRAFT_443385	D8SUX9	SELML
5	Ole e 1_5A1	ARALYDRAFT_481639	D7LGX1	ARALY
5	Ole e 1_5A2	At2g27385	O6NLE8	ARATH
5	Ole e 1_5B1	-	C6SVU8	SOYBN
5	Ole e 1_5B2	-	C6TUF4	SOYBN
5	Ole e 1_5C1	-	C6SZO1	SOYBN
5	Ole e 1_5D1	POPTRDRAFT_821590	A9PI57	POPTR
5	Ole e 1_5D2	RCOM_1281970	B9SCW4	RICCO
5	Ole e 1_5D3	VITISV_031997	A5BY12	VITVI
5	Ole e 1_5E1	At5g22430	Q9FMQ8	ARATH
5	Ole e 1_5E2	-	Q8L9I4	ARATH
5	Ole e 1_5E3	ARALYDRAFT_351256	D7MOX5	ARALY
6	Ole e 1_6A1	-	B6TL01	MAIZE
6	Ole e 1_6A2	-	B4FQB6	MAIZE

Ole e 1 Family	Revised annotation	Previous annotation	GenBank Accession number	Source
6	Ole e 1_6B1	-	B6TXH9	MAIZE
6	Ole e 1_6C1	Sb04g021840	C5XTZ8	SORBI
6	Ole e 1_6D1	B1136M02.23	Q6EPW8	ORYSJ
7	Ole e 1_7A1	SELMODRAFT_405039	D8QY68	SELML
7	Ole e 1_7A2	SELMODRAFT_414879	D8RTV5	SELML
8	Ole e 1_8A1	SELMODRAFT_448129	D8T4Z3	SELML
8	Ole e 1_8B1	SELMODRAFT_409805	D8RCI1	SELML
8	Ole e 1_8C1	SELMODRAFT_448128	D8T4Z1	SELML
9	Ole e 1_9A1	At2g21140	Q8SXP9	ARATH
9	Ole e 1_9A2	Proline-rich protein 2	Q9M7P0	ARATH
9	Ole e 1_9A3	ARALYDRAFT_900573	D7LL03	ARALY
9	Ole e 1_9B1	Extensin-like protein	Q8M6T6	ARATH
9	Ole e 1_9B2	Proline-rich protein 4	Q8M7N8	ARATH
9	Ole e 1_9B3	AT4g18770/T9A14_50	Q8T0I5	ARATH
9	Ole e 1_9B4	ARALYDRAFT_490641	D7MFN2	ARALY
10	Ole e 1_10A1	Proline-rich protein	Q8M6T7	NICGL
10	Ole e 1_10B1	VIT_00024081001	D7U5A0	VITVI
10	Ole e 1_10B2	POPTRDRAFT_200888	B9HRA8	POPTR
10	Ole e 1_10C2	POPTRDRAFT_195615	B9H154	POPTR
10	Ole e 1_10D2	RCOM_0860490	B8BTC5	RICCO
11	Ole e 1_11A1	Proline-rich protein	Q82068	SOLTU
12	Ole e 1_12A1	VITISV_029841	A5BQP2	VITVI
13	Ole e 1_13A1	VITISV_029838	A5BQP1	VITVI
13	Ole e 1_13A2	VITISV_029837	A5BQP0	VITVI
13	Ole e 1_13B1	VIT_00024076001	D7U587	VITVI
14	Ole e 1_14A1	proline-rich protein	Q83WF4	ORYSA
14	Ole e 1_14A2	Os10g0148100	Q93WL9	ORYSA
14	Ole e 1_14A3	proline-rich protein	Q93WL9	ORYSA
14	Ole e 1_14A4	Os10g0149000	Q7XGT3	ORYSJ
14	Ole e 1_14A5	proline-rich protein	Q84H18	ORYSA
14	Ole e 1_14A6	Os10g0149200	Q7XGT1	ORYSJ
14	Ole e 1_14A7	OsJ_30733	A3C7J8	ORYSJ
14	Ole e 1_14A8	OsI_12924	A2XKE8	ORYSI
14	Ole e 1_14A9	OsI_12923	A2XKE7	ORYSI
14	Ole e 1_14A10	OsJ_12921	B8AP23	ORYSI
14	Ole e 1_14A11	OSJNBa0031A07.6	Q84H17	ORYSA
14	Ole e 1_14A12	OsJ_30737	A3C7K3	ORYSJ
14	Ole e 1_14A13	Os10g0149400	Q7XGT0	ORYSJ
14	Ole e 1_14A14	OsJ_30734	A3C2K0	ORYSJ

Table 1. The Ole e 1 protein superfamily: new and unified nomenclature. ARATH: *Arabidopsis thaliana*; ARALY: *Arabidopsis lyrata*; BETPN: *Betula pendula*; BRAOL: *Brassica oleracea*; BRARP: *Brassica rapa*; CAPAN: *Capsicum annuum*; CARAS: *Cardaminopsis arenosa*; CHE1: *Chenopodium album*; CROSA: *Crocus sativus*; DAUCA: *Daucus carota*; EUPPU: *Euphorbia pulcherrima*; FRAEX: *Fraxinus excelsior*; GOSBA: *Gossypium barbadense*; GOSHE: *Gossypium herbaceum*; GOSHI: *Gossypium hirsutum*; GOSKI: *Gossypioides kirkii*; HYAOR: *Hyacinthus orientalis*; LigVu: *Ligustrum vulgare*; LILLO: *Lilium longiflorum*; LOLPE : *Lolium perenne*; MAIZE: *Zea mays*; MEDTR: *Medicago truncatula*; NICAL: *Nicotiana alata*; NICGL: *Nicotiana glauca*; NicLa: *Vitis pseudoreticulata*; OleEu: *Olea europaea*; ORYSI: *Oryza sativa*; PETCR: *Petroselinum crispum*; PETHY: *Petunia hybrida*; PHAVU: *Phaseolus vulgaris*; PHEPR : *Phleum pratense*; PHYPA: *Physcomitrella patens*; PICSI: *Picea sitchensis*; PLALA: *Platanus lanceolata*; POPTR: *Populus trichocarpa*; RICCO: *Ricinus communis*; SALKA: *Salsola kali*; SAMNI: *Sambucus nigra*; SELML: *Selaginella moellendorffii*; SOLLI: *Solanum lycopersicum*; SOLTU: *Solanum tuberosum*; SORBI: *Sorgum bicolor*; SOYBN: *Glycine max*; TOBAC: *Nicotiana tabacum*; TRISU: *Trifolium subterraneum*; VITVI: *Vitis vinifera*; 9ROSI: *Cleome spinosa*; (-): uncharacterized.

Table 1. (continued). The Ole e 1 protein superfamily: new and unified nomenclature.

Table 1. (continued). The Ole e 1 protein superfamily: new and unified nomenclature.

48	Ole e 1_48H8	-	B4FKQ2	MAIZE
48	Ole e 1_48H9	Pollen-specific protein C13	B6T720	MAIZE
48	Ole e 1_48I1	Putative pollen specific prot.C13	Q8RU50	ORYSJ
48	Ole e 1_48I2	Os10g0371000 protein	Q0IYJ9	ORYSJ
48	Ole e 1_48I3	-	A2Z6J6	ORYSI
48	Ole e 1_48I4	Pollen-specific protein C13	B6SJ40	MAIZE
48	Ole e 1_48I5	Pollen-specific protein C13	B6T594	MAIZE
48	Ole e 1_48I8	Sb0012s014630	C6JRR2	SORBI
48	Ole e 1_48I7	Pollen-specific protein	Q877C4	HYAOR
48	Ole e 1_48J1	Major pollen allergen Lol p 11	Q7M1X5	LOLPR
48	Ole e 1_48J2	Pollen allergen Phl p 11	Q8H6L7	PHLPR
48	Ole e 1_48J3	Sb03g001020	C5XKB6	SORBI
48	Ole e 1_48J4	Pollen allergen Phl p 11	B6T2Z8	MAIZE
48	Ole e 1_48J5	-	A2YE17	ORYSI
48	Ole e 1_48J6	Os06g0556800 protein	Q5Z7I0	ORYSJ
48	Ole e 1_48K1	Sb08g007260	C5YUU2	SORBI
48	Ole e 1_48K2	-	B4FCC1	MAIZE
48	Ole e 1_48K3	Pollen allergen Phl p 11	B6TN95	MAIZE
48	Ole e 1_48L1	-	B8BEU8	ORYSI
48	Ole e 1_48L2	-	A3C1R9	ORYSI
48	Ole e 1_48L3	Putative Pollen specific protein C13	Q850Z4	ORYSJ
48	Ole e 1_48L4	Os08g0572800 protein	Q0JZF0	ORYSJ
48	Ole e 1_48L5	Os07g0590500 protein	Q6ZLH6	ORYSJ
48	Ole e 1_48L6	-	A3BLQ7	ORYSJ
48	Ole e 1_48L7	-	A2YNB3	ORYSI
48	Ole e 1_48L8	Sb07g012930	C5X6P8	SORBI
48	Ole e 1_49A1	-	C6T335	SOYBN
50	Ole e 1_50A1	Pollen ole e 1 allergen	D7KV67	ARALY
50	Ole e 1_50A2	Pollen ole e 1 allergen	D7KV68	ARALY
50	Ole e 1_50A3	P-glycerate mutase 1 like prot.	Q8LD45	ARATH
50	Ole e 1_50A4	-	Q8H7B9	ARATH
50	Ole e 1_50A5	Pollen specific protein	Q42043	ARATH
51	Ole e 1_51A1	F28K19.26 protein	Q9SGZ6	ARATH
52	Ole e 1_52A1	At1g29140	O64WD6	ARATH
52	Ole e 1_52A2	F28N24.16 protein	Q8LP44	ARATH
52	Ole e 1_52B1	At5g45880	Q698L9	ARATH
52	Ole e 1_52B2	Ole e 1-like protein	Q8L9P9	ARATH
52	Ole e 1_52B3	-	D7MSQ6	ARALY
52	Ole e 1_52B4	At4g18596	Q9NMJ2	ARATH
52	Ole e 1_52B5	Pollen ole e 1 allergen	D7MCQ3	ARALY
52	Ole e 1_52B6	Ole e 1-like protein	Q9FJ48	ARATH

52	Ole e 1_52C1	Ole e 1-like protein	O49813	BETPN
52	Ole e 1_52D1	Pollen allergen Che a 1	B5S8A8	RICCO
52	Ole e 1_52E1	PN40024	D7TJL1	VITVI
52	Ole e 1_52F1	-	C6TL27	SOYBN
52	Ole e 1_52F2	-	B7FGN2	MEDTR
52	Ole e 1_52G1	Pollen allergen Che a 1	Q8LGR0	CHE1
52	Ole e 1_52G2	Pollen allergen Cro s 1	Q29W25	CROSA
52	Ole e 1_52H1	Sal k 4	E2D0Z8	SALKA
52	Ole e 1_52I1	-	B5N635	POPTR
52	Ole e 1_52I2	-	B9PUZ0	POPTR
52	Ole e 1_52I3	-	B9I1V1	POPTR
52	Ole e 1_52J1	Anther-specific prot. LAT52	B9SBK9	RICCO
52	Ole e 1_52K1	AS1	D7R0W3	GOSHI
52	Ole e 1_52L1	Anther-specific prot. LAT52	P13447	SOLLC
53	Ole e 1_53A1	-	D7KDQ6	ARALY
54	Ole e 1_54A1	Pollen-specific protein - like	O49527	ARATH
55	Ole e 1_55A1	Putative Ole e 1-like protein	A3F4A8	NicLa
56	Ole e 1_56A1	Major pollen allergen Pla l 1	P82242	PLALA
57	Ole e 1_57A1	Allergen Fra a 1.0101	Q7XAV4	FRAEX
57	Ole e 1_57A2	Fra e 1.0102 major allergen	Q5EXJ6	FRAEX
57	Ole e 1_57A3	Major pollen allergen Lig v 1	O82015	LigVu
57	Ole e 1_57A4	Ole e 1 olive pollen allergen	X76387	OleEu
57	Ole e 1_57A5	Ole e 1 olive pollen allergen	AF532765	OleEu
57	Ole e 1_57A6	Ole e 1 olive pollen allergen	AF532766	OleEu
57	Ole e 1_57A7	Ole e 1 olive pollen allergen	AF532767	OleEu
57	Ole e 1_57A8	Ole e 1 olive pollen allergen	X76306	OleEu
57	Ole e 1_57A9	Ole e 1 olive pollen allergen	P19863	OleEu
57	Ole e 1_57A10	Ole e 1 olive pollen allergen	Ole e 1 Edman	OleEu
57	Ole e 1_57A11	Ole e 1 olive pollen allergen	X76385	OleEu
57	Ole e 1_57A12	Ole e 1 olive pollen allergen	Y12428	OleEu
57	Ole e 1_57A13	Ole e 1 olive pollen allergen	Y12427	OleEu
57	Ole e 1_57A14	Ole e 1 olive pollen allergen	AF500908	OleEu
57	Ole e 1_57A15	Ole e 1 olive pollen allergen	AF516277	OleEu
57	Ole e 1_57A16	Ole e 1 olive pollen allergen	AF516278	OleEu
57	Ole e 1_57A17	Ole e 1 olive pollen allergen	AF515289	OleEu
57	Ole e 1_57A18	Ole e 1 olive pollen allergen	AF515279	OleEu
57	Ole e 1_57A19	Ole e 1 olive pollen allergen	AF515281	OleEu
57	Ole e 1_57A20	Ole e 1 olive pollen allergen	AF532755	OleEu
57	Ole e 1_57A21	Ole e 1 olive pollen allergen	AF532756	OleEu
57	Ole e 1_57A22	Ole e 1 olive pollen allergen	AF532757	OleEu
57	Ole e 1_57A23	Ole e 1 olive pollen allergen	AF532760	OleEu

57	Ole e 1_57A24	Ole e 1 olive pollen allergen	AF532753	OleEu
57	Ole e 1_57A25	Ole e 1 olive pollen allergen	AF532754	OleEu
57	Ole e 1_57A26	Ole e 1 olive pollen allergen	AY137467	OleEu
57	Ole e 1_57A28	Ole e 1 olive pollen allergen	AY137468	OleEu
57	Ole e 1_57A29	Ole e 1 olive pollen allergen	AY137469	OleEu
57	Ole e 1_57A30	Ole e 1 olive pollen allergen	575756	OleEu
57	Ole e 1_57A31	Ole e 1 olive pollen allergen	Y12426	OleEu
57	Ole e 1_57A32	Ole e 1 olive pollen allergen	AF532758	OleEu
57	Ole e 1_57A33	Ole e 1 olive pollen allergen	AF532761	OleEu
57	Ole e 1_57A34	Ole e 1 olive pollen allergen	AF532762	OleEu
57	Ole e 1_57A35	Ole e 1 olive pollen allergen	AF532759	OleEu
57	Ole e 1_57A36	Ole e 1 olive pollen allergen	AF532764	OleEu
57	Ole e 1_57A37	Ole e 1 olive pollen allergen	AF532763	OleEu
57	Ole e 1_57A38	Ole e 1 olive pollen allergen	AY189880	OleEu
57	Ole e 1_57A39	Ole e 1 olive pollen allergen	AY159881	OleEu
57	Ole e 1_57A40	Allergen Fra e 1	Q6U740	FraEx
57	Ole e 1_57A41	Ole e 1 olive pollen allergen	X76541	OleEu
57	Ole e 1_57A42	Ole e 1 olive pollen allergen	X76540	OleEu
57	Ole e 1_57A43	Ole e 1 olive pollen allergen	X76539	OleEu
58	Ole e 1_58A1	-	B7FNF5	MEDTR
58	Ole e 1_58A2	-	B7FNF3	MEDTR
58	Ole e 1_58B1	-	C6SYE3	SOYBN
59	Ole e 1_59A1	Extensin-like protein	A9NMR9	PICSI
59	Ole e 1_59A2	-	A9NPL2	PICSI
59	Ole e 1_59A3	Extensin-like protein	E0ZE82	PICSI
59	Ole e 1_59A4	Extensin-like protein	E0ZE80	PICSI
59	Ole e 1_59A5	Extensin-like protein	E0ZE78	PICSI
60	Ole e 1_60A1	AT1g27100/T7N9_15	Q94EJ3	ARATH
60	Ole e 1_60A2	-	Q8L972	ARATH
60	Ole e 1_60A3	Pollen ole e 1 allergen	D7MUX1	ARALY
60	Ole e 1_60A4	-	C6SVQ8	SOYBN
60	Ole e 1_60A5	-	C6T474	SOYBN
60	Ole e 1_60A6	-	A9P8A0	POPTR
60	Ole e 1_60A7	-	A9PFL1	POPTR
60	Ole e 1_60A8	-	B9SAF6	RICCO
60	Ole e 1_60A9	PN40024	D7T895	VITVI
60	Ole e 1_60A10	-	A2X417	ORYSI
60	Ole e 1_60A11	Os02g0317800 protein	Q6Z841	ORYSJ
60	Ole e 1_60A12	Sb07g009530	C5YJW7	SORBI
60	Ole e 1_60A13	-	B6T3A4	MAIZE
60	Ole e 1_60B1	-	B6LRF7	PICSI
61	Ole e 1_61A1	-	Q8RWG5	ARATH

61	Ole e 1_61A2	At2g16630	Q8BLF4	ARATH
61	Ole e 1_61A3	-	D7LFM3	ARALY
61	Ole e 1_61B1	-	B9N159	POPTR
61	Ole e 1_61B2	-	A9PG40	POPTR
61	Ole e 1_61B3	-	B9SQJ5	RICCO
61	Ole e 1_61B4	-	D7U593	VITVI
62	Ole e 1_62A1	PN40024	A2WZR9	ORYSI
62	Ole e 1_62A2	-	A3A255	ORYSJ
62	Ole e 1_62B1	-	B6TF27	MAIZE
62	Ole e 1_62B2	-	B4FZU6	MAIZE
62	Ole e 1_62B3	-	B6TRI7	MAIZE
63	Ole e 1_63A1	-	B6UFQ0	MAIZE
64	Ole e 1_64A1	-	A9RQ15	PHYPA
64	Ole e 1_64B1	-	A9SHJ0	PHYPA
65	Ole e 1_65A1	-	D8TAV8	SELML
65	Ole e 1_65A2	-	D8TDP3	SELML
66	Ole e 1_66A1	-	D7MY20	ARALY
66	Ole e 1_66A2	proline-rich glycoprotein	O64586	ARATH
66	Ole e 1_66A3	-	D7LH37	ARALY
67	Ole e 1_67A1	-	D7LGE7	ARALY
67	Ole e 1_67A2	At2g33790	P93013	ARATH
67	Ole e 1_67B1	At1g28290	Q9FZA2	ARATH
67	Ole e 1_67B2	proline-rich protein	Q6WP47	ARATH
67	Ole e 1_67B3	-	D7KCU8	ARALY
67	Ole e 1_67C1	HyPRP1	Q6PIW3	GOSHI
67	Ole e 1_67D1	Arabinogalactan protein	C6YQU7	GOSHI
68	Ole e 1_68A1	-	C6TLD2	SOYBN
68	Ole e 1_68A2	proline-rich protein	Q41122	PHAVU
68	Ole e 1_68B1	-	B7F59	MEDTR
68	Ole e 1_68C1	-	A9PAW5	POPTR
68	Ole e 1_68C2	-	A9PA42	POPTR
68	Ole e 1_68D1	-	B9N3G7	POPTR
68	Ole e 1_68D2	-	B9N2T9	POPTR
68	Ole e 1_68D3	-	A9P845	POPTR
68	Ole e 1_68E1	Structural constituent of cell wall	B9RBC9	RICCO
68	Ole e 1_68F1	PN40024	D7TGB4	VITVI
68	Ole e 1_69O1	Arabinogalactan protein	Q9FSW6	DAUCA
69	Ole e 1_69A1	hybrid proline-rich protein PRP1	Q9XES6	TRISU
70	Ole e 1_70A1	Proline-rich protein	Q07894	NICAL
70	Ole e 1_70A2	-	CULL53	PETHY
70	Ole e 1_70A3	Proline-rich protein 1	Q6ONA3	CAPAN

71	Ole e 1_71A1	Pistil extensin-like protein	Q40385	NICAL
71	Ole e 1_71A2	Pistil-specific extensin-like prot.	Q03211	PEXLP
71	Ole e 1_71B1	Pistil extensin like protein	Q40549	TOBAC
71	Ole e 1_71B1	Pistil extensin like protein	Q40552	TOBAC
72	Ole e 1_72A1	120 kDa style glycoprotein	Q49986	NICAL
72	Ole e 1_72A2	120 kDa pistil extensin-like prot.	Q49W28	NicLa
72	Ole e 1_72A3	120 kDa pistil extensin-like prot.	Q49W29	NicLa
72	Ole e 1_72A4	120 kDa pistil extensin-like prot.	Q49W27	NicLa
72	Ole e 1_72A5	120 kDa pistil extensin-like prot.	Q49W32	NicLa
72	Ole e 1_72A6	120 kDa pistil extensin-like prot.	Q49W33	NICPL
72	Ole e 1_72A7	120 kDa pistil extensin-like prot.	Q49W34	TOBAC
73	Ole e 1_73A1	120 kDa pistil extensin-like prot.	Q49W30	NicLa
74	Ole e 1_74A1	Pollen ole e 1 allergen	D7MA28	ARALY
74	Ole e 1_74A2	At4g02770	QB1417	ARATH
75	Ole e 1_75A1	-	C6T5T7	SOYBN
75	Ole e 1_75A2	Drought resistance protein	E0A235	SOYBN
75	Ole e 1_75A3	-	C6T325	SOYBN
75	Ole e 1_75B1	-	B9N9P3	POPTR
75	Ole e 1_75B2	-	B9N9P2	POPTR
75	Ole e 1_75C1	-	B9MX40	POPTR
75	Ole e 1_75C2	-	B9P957	POPTR
75	Ole e 1_75D1	-	B9GSD2	POPTR
76	Ole e 1_76A1	-	B9SAV5	RICCO
76	Ole e 1_76B1	Structural constituent cell wall	B9SAV4	RICCO
77	Ole e 1_77A1	-	B9GSD1	POPTR
77	Ole e 1_77A2	Structural constituent cell wall	B9SAV3	RICCO
77	Ole e 1_77B1	PN40024	D7U2C5	VITVI
77	Ole e 1_77B2	-	A5B1Z7	VITVI
77	Ole e 1_77C1	PN40024	D7U2C3	VITVI
77	Ole e 1_77C2	-	A5B1Z5	VITVI
77	Ole e 1_77D1	Pollen ole e 1 allergen	D7LGP2	ARALY
77	Ole e 1_77D2	At2g47540	Q7Z57	ARATH
78	Ole e 1_78A1	-	A5B1Z6	VITVI
78	Ole e 1_78A2	PN40024	D7U2C4	VITVI
79	Ole e 1_79A1	-	D7KQ21	ARALY
79	Ole e 1_79A2	-	D7KQ74	ARALY
79	Ole e 1_79A3	Proline-rich protein 1	Q9FZ35	ARATH
79	Ole e 1_79A4	Proline-rich protein 1	Q9M7P1	ARATH
79	Ole e 1_79A5	Proline-rich protein	Q9LZJ7	ARATH
79	Ole e 1_79A6	Proline-rich protein 3	Q9M7N9	ARATH
79	Ole e 1_79A7	-	D7LT86	ARALY
80	Ole e 1_80A1	-	D7LGP0	ARALY

80	Ole e 1_80A2	-	Q22258	ARATH
81	Ole e 1_81A1	-	D8TCE6	SELML
81	Ole e 1_81A2	-	D8TF48	SELML
81	Ole e 1_81B1	-	D8TCE7	SELML
82	Ole e 1_82A1	-	D8TCF8	SELML
83	Ole e 1_83A1	Pollen ole e 1 allergen	D7LYX8	ARALY
83	Ole e 1_83A2	At3g05500	Q9FFG5	ARATH
83	Ole e 1_83B1	-	B9HHU1	POPTR
83	Ole e 1_83B2	-	B9HSK7	POPTR
83	Ole e 1_83B3	-	B9SQR8	RICCO
83	Ole e 1_83B4	PN40024	D7T4L1	VITVI
83	Ole e 1_83B5	-	A5C9V2	VITVI
84	Ole e 1_84A1	-	A2WL03	ORYSJ
84	Ole e 1_84A2	-	Q8LJN2	ORYSJ
84	Ole e 1_84A3	-	A2WL01	ORYSJ
84	Ole e 1_84A4	-	B9ETU7	ORYSJ
84	Ole e 1_84A5	B1189A09.32	Q5VR52	ORYSJ
84	Ole e 1_84A6	-	A2WL00	ORYSJ
84	Ole e 1_84A7	-	A2WL05	ORYSJ
84	Ole e 1_84A8	-	A2WL08	ORYSJ
84	Ole e 1_84A9	-	Q6LJM9	ORYSJ
84	Ole e 1_84A10	-	A2WL07	ORYSJ
84	Ole e 1_84A11	-	Q6LJM8	ORYSJ
84	Ole e 1_84A13	-	A2WL04	ORYSJ
84	Ole e 1_84A14	-	Q8LJN3	ORYSJ
84	Ole e 1_84A15	B1189A09.34	A2ZPK8	ORYSJ
84	Ole e 1_84B1	-	Q5VR30	ORYSJ
84	Ole e 1_84C1	-	A2WL02	ORYSJ
84	Ole e 1_84C2	-	A2WL12	ORYSJ
84	Ole e 1_84C3	-	Q8LJM5	ORYSJ
84	Ole e 1_84C4	-	B9ADE9	ORYSJ
84	Ole e 1_84C5	B1189A09.42	A2ZPL3	ORYSJ
84	Ole e 1_84C6	-	Q5VR18	ORYSJ
84	Ole e 1_84C7	-	A2WL11	ORYSJ
84	Ole e 1_84D1	B1189A09.38	Q6LJM6	ORYSJ
84	Ole e 1_84E1	-	Q5VR19	ORYSJ
84	Ole e 1_84F1	-	B9ET08	ORYSJ
84	Ole e 1_84F2	B1189A09.45	A2ZPL6	ORYSJ
84	Ole e 1_84G1	-	Q5VR17	ORYSJ
84	Ole e 1_84G2	Sb03g005060	A2WL13	ORYSJ
85	Ole e 1_85A1	-	C5XP48	SORBI
			Q8SHH8	SELML

85	Ole e 1_85B1	-	D8T5S3	SELML
86	Ole e 1_86A1	-	D8SHH9	SELML
87	Ole e 1_87A1	-	D8T5S4	SELML
88	Ole e 1_88A1	-	D8R4E3	SELML
88	Ole e 1_88A2	-	D8RRK7	SELML
89	Ole e 1_89A1	Pollen ole e 1 allergen	D7L8E4	ARALY
89	Ole e 1_89A2	-	Q3EBA2	ARATH
89	Ole e 1_89B1	-	B9MTK8	POPTR
89	Ole e 1_89B2	-	B9S9B7	RICCO
89	Ole e 1_89C1	-	C6T3U0	SOYBN
90	Ole e 1_90A1	PN40024	D7T1W8	VITVI
90	Ole e 1_90A2	PN40024	D7T1X1	VITVI
91	Ole e 1_91A1	-	A2Z9X7	ORYSI
91	Ole e 1_91A2	Os10g0546100	Q9AV31	ORYSJ
91	Ole e 1_91B1	Sb01g030890	C5WTH1	SORBI
92	Ole e 1_92A1	-	D7LH6A	ARALY
92	Ole e 1_92A2	At2g41400	Q6DBF8	ARATH
92	Ole e 1_92A3	RAFL22-83-M12	Q67ZJ7	ARATH
92	Ole e 1_92B1	At2g41400	Q9ZVC5	ARATH
92	Ole e 1_92B2	At2g41390	Q9ZVC4	ARATH
93	Ole e 1_93A1	At5g05020	Q9FF72	ARATH
94	Ole e 1_94A1	-	D7KW90	ARALY
95	Ole e 1_95A1	At3g16660	Q8GYY6	ARATH
95	Ole e 1_95A2	MGL6	Q8LUR8	ARATH
95	Ole e 1_95A3	-	D7L657	ARALY
95	Ole e 1_95B1	-	D7L658	ARALY
95	Ole e 1_95B2	AT3G16670/MGL6_12	Q8LUR5	ARATH
95	Ole e 1_95C1	-	B9HXT5	POPTR
95	Ole e 1_95C2	PN40024	D7SYD6	VITVI
95	Ole e 1_95C3	Phytoplanin	B9RT72	RICCO
95	Ole e 1_95D1	-	C6TK65	SOYBN
95	Ole e 1_95E1	Phytoplanin	Q56S59	PHYLL
95	Ole e 1_95F1	Phytoplanin	Q1PCF2	TOBAC
96	Ole e 1_96A1	-	B8APE4	ORYSI
96	Ole e 1_96A2	-	B9F8E1	ORYSJ
96	Ole e 1_96B1	-	A2YPV4	ORYSI
96	Ole e 1_96B2	Os07g0874400 protein	Q6ZDW8	ORYSJ
96	Ole e 1_96B3	Sb07g042710	C5X5I9	SORBI
96	Ole e 1_96B4	-	B4FF58	MAIZE
96	Ole e 1_96B5	Sb01g035R10	C5XBQ0	SORBI
97	Ole e 1_97A1	-	A2YPV5	ORYSI
97	Ole e 1_97A2	Os07g0674500	Q6ZDW7	ORYSJ

97	Ole e 1_97A3	Sb02g042740	C5X5J0	SORBI
98	Ole e 1_98A1	-	A2XGJ7	ORYSI
98	Ole e 1_98A2	-	Q10LN4	ORYSJ
98	Ole e 1_98B1	-	A2XGJ8	ORYSI
98	Ole e 1_98C1	Os03g0342100	Q10LN2	ORYSJ
98	Ole e 1_98C1	Sb01g035830	C5X0Q2	SORBI
99	Ole e 1_99A1	-	B6UHT3	MAIZE
100	Ole e 1_100A1	-	A2WUN2	ORYSI
100	Ole e 1_100A2	Os01g0725000	Q68158	ORYSJ
100	Ole e 1_100A3	-	A2ZXF1	ORYSJ
100	Ole e 1_100A4	Sb03g033350	C5XIF5	SORBI
100	Ole e 1_100B1	Sb03g033360	C5XIF6	SORBI
100	Ole e 1_100C1	OJ1131_E09.17	Q75K53	ORYSJ
100	Ole e 1_100C2	-	B9FH39	ORYSJ
100	Ole e 1_100C3	-	A2YET6	ORYSI
100	Ole e 1_100C4	Os05g0531400 protein	Q6DGH6	ORYSJ
100	Ole e 1_100D1	Sb09g026510	C5YUF6	SORBI
100	Ole e 1_100D2	Arabinogalactan protein	B6SLV3	MAIZE
101	Ole e 1_101A1	-	A2YET4	ORYSI
101	Ole e 1_101A2	Os05g0531200 protein	Q75K55	ORYSJ
101	Ole e 1_101B1	Pistil-specific extensin-like protein	B6UHM8	MAIZE
101	Ole e 1_101B2	-	B4FGH8	MAIZE
101	Ole e 1_101B3	Sb09g028500	C5Z1H9	SORBI
102	Ole e 1_102A1	-	A2WUN3	ORYSI
102	Ole e 1_102A2	Os01g0726100 protein	Q8S154	ORYSJ
102	Ole e 1_102B1	Sb03g033370	C5XIF7	SORBI
102	Ole e 1_102B2	Pistil-specific extensin-like prot.	B6UHE3	MAIZE
103	Ole e 1_103A1	-	D8RZL9	SELML
104	Ole e 1_104A1	-	B4FMQ8	MAIZE
105	Ole e 1_105A1	-	D8RCl2	SELML
106	Ole e 1_106A1	-	D8T995	SELML
107	Ole e 1_107A1	-	A9SC13	PHYPA
108	Ole e 1_108A1	-	D8QY67	SELML
109	Ole e 1_109A1	-	D8S5V6	SELML

Table 1. (continued). The Ole e 1 protein superfamily: new and unified nomenclature.

3.2 Phylogenetic analysis of the extended Ole e 1 protein families

A member of each retrieved full-length Ole e 1 sequences family was aligned to determine phylogenetic relationships within the Ole e 1 extended family. A phylogenetic tree of the Ole e 1 extended sequences is depicted in Figure 1.

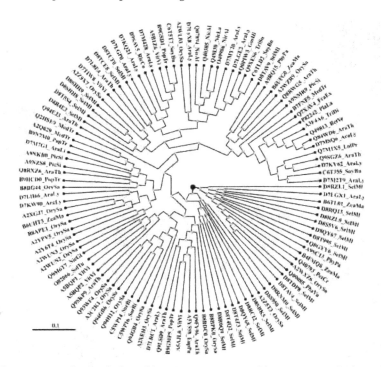

Fig. 1. Phylogenetic analysis of plant Ole e 1 proteins. Neighbour-Joining (NJ) method was used to perform a phylogenetic analysis of Ole e 1 proteins from 109 families. One representative sequence of each family was used, based in its higher consensus ability. Plant species analyzed included *Arabidopsis*, poplar, rice, spikemoss, tobacco, maize, potato, grape, *Sorghum*, kidney bean, barrel medic, *Pinus*, poinsettia, perennial ryegrass, soybean, white birch, ash, *Platanus*, *Physcomitrella*, cotton, subterranean clover, Persian tobacco and castor bean.

The phylogenetic tree shows that the 109 Ole e 1 extended families, although highly divergent, are split into two clades. The smaller clade was integrated by a few species like *Selaginella moellendorffii*, *Arabidopsis* and maize among others. The second clade included the majority of the Ole e 1 family proteins, clustering together almost all the biological functions (Figure 1). Numerous branches aroused from this clade.

3.3 Ole e 1 protein superfamilies: Structural and conformational variability

The crystallographic structural coordinates of relatively few proteins of the Ole e 1 family have been deposited in the Protein Database (PDB) up to date. To our knowledge, detailed comparative studies of the structural and conformational features of members of the Ole e 1 extended protein families have not been performed in higher plants. Using computational modelling analysis, we have determined and modelled the molecular-structural features of selected members of the Ole e 1 extended families. A first overview of the generated models (Figure 2) indicated a relatively high level of similitude.

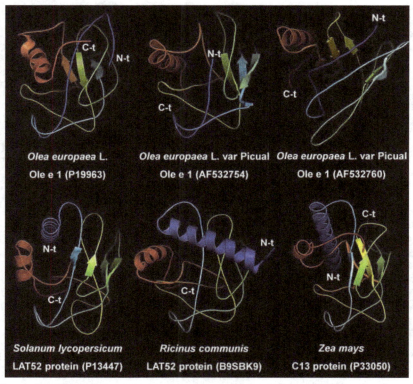

Fig. 2. Three-dimensional structure analysis of selected members of Ole e 1 family proteins. The model proteins are depicted as cartoon diagrams. The secondary elements of the crystallographic structures are rainbow coloured, with N-terminus in blue, and C-terminus in red.

However, a more detailed analysis allowed identifying certain differences in the generated models, particularly consisting in 2D structural features. These differences can be distinguished even between very close proteins like P19963, AF532754 and AF532760 (Ole e 1_57A9, Ole e 1_57A25 and Ole e 1_57A23 with the new nomenclature), corresponding to the olive pollen major allergen cloned from different varietal sources or even to different clones of the same cultivar (Figure 2). The differences become higher when models of the same protein obtained from different plant species are compared. This is the case of P13447 and B9SBK9 (Ole e 1_52L1 and Ole e 1_52J1), which correspond to the LAT52 gene product in tomato and *Ricinus*, respectively (Figure 2). Divergences are even more obvious between the models indicated above and that of a P33050 (Ole e 1_48H6), a different member of the Ole e 1 superfamily corresponding to a pollen protein from maize (C13 protein) (Figure 2).

4. Discussion

Research as regard to the proteins of the Ole e 1 family has been carried out steadily since its definition. At present, many genes from the allergen Ole e 1 family of proteins have been characterized, and data are available concerning the sequence, structure, expression and biological function (e.g. extensin-like proteins constituting part of the cell wall). However, and as depicted in this chapter, the precise identification of more than half members of this family remains uncompleted. Up to now, Ole e 1 and Ole e 1-like genes are deposited into the databases, many of them with repetitive or arbitrary naming system by authors. This nomenclature includes a variety of generic names, such as Ole e 1 major olive pollen allergen, putative Ole e 1-like protein, anther-specific Ole e 1-like protein, and others depending of the protein location in the chromosome, e.g. At3g26960, Os09g0508200, or simply giving a random name e.g. P1 clone: MOJ10. For those members of the Ole e 1 family which have been recognized like allergens, a more sustainable and precise nomenclature has been built, by following the recommendations of the International Union of Immunological Societies (IUIS) (http://www.allergen.org/). However, these allergenic proteins only represent a part of the members of the Ole e 1 family, and this nomenclature still does not display the relationships among these proteins. In several cases, it is still common for researchers to use different names for the same allergen. Allergen biochemistry is now entering a new time of structural biology and proteomics that will require sophisticated tools for data processing and bioinformatics, and might require further definition of the nomenclature. Increasingly, the wealth of structural information is enabling the biologic function of allergens to be established and the assignment of allergen function to diverse protein families. Therefore, the arbitrary nomenclature currently in use is not sustainable for adequate comparative mega-functional genomics studies, especially as the number of Ole e 1 genes has increased steadily and will continue with this upward trend with the completion of the sequencing projects corresponding to more plant genomes.

The implementation of modifications in the nomenclature as proposed here may assist further developments of allergy understanding and new clinical approaches. As an example, nomenclature and structural biology have been proposed to play a crucial role in defining allergens for research studies and for the development of new clinical products [Chapman et al. 2007]. Sequence comparisons and assignments to protein families provide a molecular basis for clinical cross-reactions between food, pollen, and latex allergens that give rise to oral allergy syndromes [Wagner et al. 2002, Scheiner et al. 2004, van Ree 2004]. For food and pollen allergens, intrinsic protein structure probably plays an important role in determining allergenicity by conferring, for example, heat stability or resistance to digestion in the digestive tract, e.g. storage proteins from seed/nuts or legumes [Orruño and Morgan 2011]. Interestingly, analysis of databases, e.g. pFAM shows that there are currently more than 120 molecular architectures that are responsible for eliciting IgE responses. It will be important to link nomenclature with classification of allergens into protein families and subfamilies to provide complete definition of allergens and their structure-functional relationships as part of a comprehensive bioinformatics database. The practical consequences of this approach are seen most clearly with genetically modified foods, in which sequence comparisons can be used for safety assessment of genetically modified organisms [Goodman and Tetteh 2011].

The success of our new and unified nomenclature lies in its simplicity, with genetic basis and structural-functional characterizations of the proteins, regardless of the species origin, with the possibility to further nomenclature expansion, to include as-yet-unidentified protein allergens from different sources or species: mites, insects, pollens, molds and foods. It might be also possible to include in the system engineered protein molecules, such as hypoallergens, or others being described as non-protein allergens. Allergens entered into the nomenclature could be used to develop allergen-specific diagnostics and to formulate recombinant allergen vaccines that will benefit patients [Chapman et al. 2000, Ferreira et al. 2004, Jutel et al. 2005, Sastre 2010].

The proposed system may also assist to clarify the importance of allergen polymorphism. Allergens often display numerous variants. These are proteins with typically greater than 90% sequence identity, but with enough differences in their amino acid sequences to make worth individual structural and or functional characterization and identification. This polymorphism has been deeply analyzed in mites, as their allergens present an extensive number of isoforms: 23 for Der p 1 and 13 for Der p 2 [Smith et al. 2001, Smith et al. 2001]. Furthermore, these polymorphisms might affect T-cell responses or alter antibody-binding sites. These differences can be structurally characterized to distinguish isoforms in a well-defined nomenclature system, by mean of structural-functional differentiation, helping to design allergen formulations for immunotherapy [Jutel et al. 2005, Piboonpocanun et al. 2006]. In the case of pollen allergens, Ole e 1 from olive pollen is a clear example of extreme polymorphism, both in its peptide and in its carbohydrate moieties, as demonstrated by peptide mapping and N-glycopeptide analysis [Castro et al. 2010]. Olive cultivar origin is a major cause of polymorphism for Ole e 1 pollen allergen [Hamman-Khalifa et al. 2008, Castro et al. 2010]. The olive tree has an extremely wide germplasm, with over 1200 varieties cultivated over the world [Bartolini et al. 1994]. Therefore, the number of Ole e 1 isoforms yet to be characterized in olive pollen is expected to be enormous. A similar situation is also likely to occur in many other plant species.

Overall, our developed unified nomenclature system is helpful in a quick functional prediction of any newly cloned Ole e 1 gene(s), because from the nomenclature point of view, the newly sequenced gene(s) will always be characterized/named with sequence similarity with previously characterized Ole e 1 genes/proteins, as well as a protein structure-functional characterization and comparison. The changes that have been introduced reflect into which extended family or subfamily a certain Ole e 1 protein belongs. Accordingly, the new nomenclature will have no significant impact on already published data with old/arbitrary naming system. However, we urge scientists working on Ole e 1's to adopt this new and easy nomenclature system. In this regard, we have made an effort to preserve the user friendly linkage between the old and the new designations, which we hope will help researchers to adapt the new names. As the revised nomenclature should facilitate communication and understanding within the community interested in Ole e 1 allergen proteins, we advocate that this new naming system be used in all future studies.

The classification model used here has been developed under the basis of a previously designed gene nomenclature model for male fertility restorer (RF) proteins in higher plants [Kotchoni et al. 2010]. The increasing numbers of RF genes described in the literature represented an ongoing challenge in their clear identification and logical classification which was solved using the proposed nomenclature. Undoubtedly, similar approaches could be applied to numerous protein families involving relevant levels of nomenclature heterogeneity, many of them registered in specialized databases like pFam. In the case of allergens, other numerous protein families like profilins (Ole e 2 in the case of olive pollen) prolamins, cupins, Bet v 1-related proteins etc., which are currently included in the AllFam database [Radauer et al. 2008] (http://www.meduniwien.ac.at/allergens/allfam/) could benefit of the use of similar approaches.

5. Conclusion

We propose for first time a unified naming system for Ole e 1-like genes and pseudogenes across all plant species, which accommodates the numerous sequences already deposited in several databases, offering the needed flexibility to incorporate additional Ole e 1-like proteins as they become available. Additionally, we provide an analysis of the phylogenetic relationships displayed by the members of the Ole e 1-like family and use computational protein modelling to determine structural features of selected members of this family. These data are of particular relevance for the understanding of their biological activity and allergenic cross-reactivity.

6. Acknowledgment

Support of the Spanish Ministry of Science and Innovation (ERDF-cofinanced project BFU2008-00629) and Andalusian Regional Government (ERDF-cofinanced Proyectos de Excelencia CVI5767 and AGR6274) is gratefully acknowledged.

7. References

Alché, J.D.; Castro, A.J.; Olmedilla, A.; Fernández, M.C.; Rodríguez, R.; Villalba, M. and Rodríguez-García, M.I. (1999). The major olive pollen allergen (Ole e I) shows both gametophytic and sporophytic expression during anther development, and its synthesis and storage takes place in the RER. *Journal of Cell Science*, Vol.112, pp.2501-2509

Alché, J.D.; M'rani-Alaoui, M.; Castro, A.J. and Rodríguez-García, M.I. (2004). Ole e 1, the major allergen from olive (*Olea europaea* L.) pollen, increases its expression and is released to the culture medium during in vitro germination. *Plant Cell Physiology*, Vol.45, pp.1149-1157

Altschul, S.F.; Gish, W.; Miller, W.; Myers, E.W. and Lipman, D.J. (1990). Basic local alignment search tool. Journal of Molecular Biology, Vol.215, No.3, pp.403-410

Altschul, S.F.; Madden, T.L.; Schäffer, A.A.; Zhang, J.; Zhang, Z.; Miller, W. and Lipman, D.J. (1997). Gapped BLAST and PSI-BLAST: a new generation of protein database search programs. *Nucleic Acids Research*, Vol.25, No.17, pp.3389-402

Barral, P.; Batanero, E.; Palomares, O.; Quiralte, J.; Villalba, M. and Rodríguez, R. (2004). A major allergen from pollen defines a novel family of plant proteins and shows intra- and interspecies cross-reactivity. *Journal of Immunology*, Vol.172, pp.3644-3651

Bartolini, G.; Prevost, G. and Messeri, C. (1994). Olive tree germplasm: descriptor lists of cultivated varieties in the world. *Acta Horticulturae*, Vol.365, pp.116-118

Batanero, E.; Villalba, M. and Rodríguez, R. (1994). Glycosylation site of the major allergen from olive tree. Allergenic implications of the carbohydrate moiety. *Molecular Immunology*, Vol.31, pp.31-37

Castro, A.J.; Bednarczyk, A.; Schaeffer-Reiss, C.; Rodríguez-García, M.I.; Van Dorsselaer, A.; Alché, J.D. (2010). Screening of Ole e 1 polymorphism among olive cultivars by peptide mapping and N-glycopeptide analysis. *Proteomics*, Vol. 10, No 5, pp.953-962

Chapman, M.D.; Pomés, A.; Breiteneder, H. and Ferreira, F. (2007). Nomenclature and structural biology of allergens. *Journal of Allergy and Clinical Immunology*, Vol.119, No.2, pp.414-420

Chapman, M.D.; Smith, A.M.; Vailes, L.D.; Arruda, K.; Dhanaraj, V. and Pomes, A. (2000). Recombinant allergens for diagnosis and therapy of allergic diseases. *The Journal of Allergy and Clinical Immunology*, Vol.106, pp.409-418

Chevenet, F.; Brun, C.; Banuls, A.L.; Jacq, B. and Christen, R. (2006). TreeDyn: towards dynamic graphics and annotations for analyses of trees. *BMC Bioinformatics*, Vol.7, pp.439

D'Amato, G.; Spieksma, F.T.; Liccardi, G.; Jager, S.; Russo, M.; Kontou-Fili, K.; Nikkels, H.; Wuthrich, B. and Bonini, S. (1998). Pollen-related allergy in Europe. *Allergy*, Vol.53, pp.67-78

de Castro, E.; Sigrist, C.J.A.; Gattiker, A.; Bulliard, V.; Langendijk-Genevaux, P.S.; Gasteiger, E.; Bairoch, A. and Hulo, H. (2006) ScanProsite: detection of PROSITE signature matches and ProRule-associated functional and structural residues in proteins. *Nucleic Acids Research*, Vol.34, pp.362-365

Ferreira, F.; Wallner, M. and Thalhamer, J. (2004). Customized antigens for desensitizing allergic patients. *Advances in Immunology*, Vol.84, pp.79-129

Finn, R.D.; Mistry, J.; Tate, J.; Coggill, P.; Heger, A.; Pollington, J.E.; Gavin, O.L. Gunesekaran, P.; Ceric, G. Forslund, K.; Holm, L.; Sonnhammer, E.L.; Eddy, S.R. and Bateman, A. (2010). The Pfam protein families database. *Nucleic Acids Research*, Database Issue 38, pp.D211-222

Gasteiger, E.; Gattiker, A.; Hoogland, C.; Ivanyi, I.; Appel R.D. and Bairoch A. (2003) ExPASy: the proteomics server for in-depth protein knowledge and analysis. *Nucleic Acids Research*, Vol.31, pp.3784-3788

Goodman, R.E. and Tetteh, A.O. (2011). Suggested Improvements for the Allergenicity Assessment of Genetically Modified Plants Used in Foods. *Current Allergy and Asthma Reports*, doi: 10.1007/s11882-011-0195-6

Guex, N. and Peitsch, M.C. (1997). SWISS-MODEL and the Swiss-PdbViewer: an environment for comparative protein modeling. *Electrophoresis*, Vol.18, No.15, pp.2714-2723

Hall, T.A. (1999). BioEdit: a user-friendly biological sequence alignment editor and analysis program for Windows 95/98/NT. *Nucleic Acids Symposium Series*, Vol.41, pp.95-98

Hamman-Khalifa, A.M.; Castro A.J.; Jimenez-Lopez, J.C.; Rodríguez-García, M.I. and Alché, J.D. (2008). Olive cultivar origin is a major cause of polymorphism for Ole e 1 pollen allergen. *BMC Plant Biology*, Vol.8, 10

Hanson, D.D.; Hamilton, D.S.; Travis, J.L.; Bashe, D.M. and Mascarenhas, J.P. (1998). Characterization of a pollen-specific cDNA clone from Zea mays and its expression. *Plant Cell*, Vol.1, pp.173-179

Hauser, M.; Roulias, A.; Ferreira, F. & Egger, M. (2010). Panallergens and their impact on the allergic patient. *Allergy, Asthma & Clinical Immunology*, Vol.6, pp.1-

Jutel, M.; Jaeger, L.; Suck, R.; Meyer, H.; Fiebig, H. and Cromwell, O. (2005). Allergenspecific immunotherapy with recombinant grass pollen allergens. *The Journal of Allergy and Clinical Immunology*, Vol.116, pp.608-613

Laskowski, R.A.; MacArthur, M.W.; Moss, D.S. and Thornton, J.M. (1993). PROCHECK: A program to check the stereo-chemical quality of protein structures. *Journal of Applied Crystallography*, Vol.26, pp.283-291

Lauzurica, P.; Gurbindo, C.; Maruri, N.; Galocha, B.; Diaz, R.; Gonzalez, J.; García, R. and Lahoz, C. (1988). Olive (*Olea europea*) pollen allergens—I. Immunochemical characterization by immunoblotting, CRIE and immunodetection by a monoclonal antibody. *Molecular Immunology*, Vol.25, pp.329-335

King, T.P.; Hoffman, D.; Lowenstein, H.; Marsh, D.G.; Platts-Mills, T.A. and Thomas, W. (1994). Allergen nomenclature. WHO/IUIS Allergen Nomenclature Subcommittee. *International Archives of Allergy and Immunology*, Vol. 105, pp. 224-233

Kotchoni, S.O.; Jimenez-Lopez, J.C.; Gachomo, W.E. and Seufferheld, M.J. (2010). A new and unified nomenclature for male fertility restorer (RF) proteins in higher plants. *PLoS ONE*, Vol.5, No.12, pp.e15906

Melo, F. and Feytmans, E. (1997). Novel knowledge-based mean force potential at atomic level. *Journal of Molecular Biology*, Vol.267, No.1, pp.207-222

Melo, F. and Feytmans, E. (1998). Assessing protein structures with a non-local atomic interaction energy. *Journal of Molecular Biology*, Vol.277, No.5, pp.1141-1152

Mothes, N.; Horak, F. & Valenta, R. (2004). Transition from a botanical to a molecular classification in tree pollen allergy: implications for diagnosis andtherapy. *International Archives of Allergy and Immunology*, Vol.135, pp.357-373

Orruño, E. and Morgan, M.R.A. (2011). Resistance of purified seed storage proteins from sesame (*Sesamum indicum* L.) to proteolytic digestive enzymes. *Food Chemistry*, in press

Piboonpocanun S, Malinual N, Jirapongsananuruk J, Vichyanond P, Thomas WR. (2006). Genetic polymorphisms of major house dust mite allergens. *Clinical & Experimental Allergy*, Vol.36, pp.510-516

Radauer, C.; Bublin, M.; Wagner, S.; Mari, A. and Breiteneder, H. (2008). Allergens are distributed into few protein families and possess a restricted number of biochemical functions. *Journal of Allergy and Clinical Immunology*, Vol.121, pp.847-852

Rodriguez, R.;Villalba, M.; Batanero, E.; González, E.M.; Monsalve, R.I.; Huecas, S.; Tejera, M.L. and Ledesma, A. (2002). Allergenic diversity of the olive pollen. *Allergy*, Vol.57, pp.6-16

Rodríguez, R.; Villalba, M.; Monsalve, R.I.; Batanero, E.; González, E.M.; Monsalve, R.I.; Huecas, S.; Tejera, M.L. and Ledesma, A. (2002). Allergenic diversity of the olive pollen. *Allergy*, Vol.57, pp.6-15

Salamanca, G.; Rodriguez, R. Quiralte, J.; Moreno, C.; Pascual, C.Y.; Barber, D. and Villalba, M. (2010). Pectin methylesterases of pollen tissue, a major allergen in olive tree. *FEBS Journal*, Vol.277, No.13, pp.2729-2739

Sastre, J. (2010). Molecular diagnosis in allergy. *Clinical & Experimental Allergy*, Vol.40, No.10, pp.1442-1460

Scheiner, O.; Aberer, W.; Ebner, C.; Ferreira, F.; Hoffmann-Sommergruber, K.; Hsieh, L.S.; Kraft, D.; Sowka, S.; Vanek-Krebitz, M. and Breiteneder, H. (1997). Cross-racting allergens in tree pollen and pollen-related food allergy: implications for diagnosis of specific IgE. *International Archives of Allergy and Immunology*, Vol.113, pp.105-108

Shultz, J.L.; Kurunam, D.; Shopinski, K.; Iqbal, M.J.; Kazi, S.; Zobrist, K.; Bashir, R.; Yaegashi, S.; Lavu, N.; Afzal, A.J.; Yesudas, C.R.; Kassem, M.A.; Wu, C.; Zhang, H.B.; Town, C.D.; Meksem, K. and Lightfoot, D.A. (2006). The Soybean Genome Database (SoyGD): a browser for display of duplicated, polyploid, regions and sequence tagged sites on the integrated physical and genetic maps of Glycine max. *Nucleic Acids Research*, Vol.34(suppl 1), pp.D758-D765

Sigrist, C.J.A.; Cerutti, L.; de Castro, E.; Langendijk-Genevaux, P.S.; Bulliard, V.; Bairoch, A. and Hulo, N. (2010). PROSITE, a protein domain database for functional characterization and annotation. *Nucleic Acids Research*, Vol.38 (Database issue), pp.161-166

Smith, A.S.; Benjamin, D.C.; Hozic, N.; Derewenda, U.; Smith, W.A.; Thomas, W.R.; Gafvelin, G.; van Hage-Hamsten, M. and Chapman, M.D. (2001). The molecular basis of antigenic cross-reactivity between the group 2 mite allergens. *The Journal of Allergy and Clinical Immunology*, Vol.107, pp.977-984

Smith, W.A.; Hales, B.J.; Jarnicki, A.G. and Thomas W.R. (2001). Allergens of wild house dust mites: environmental Der p 1 and Der p 2 sequence polymorphisms. *The Journal of Allergy and Clinical Immunology*, Vol.107, pp.985-992

Stratford, S.; Barne, W.; Hohorst, D.L.; Sagert, J.G.; Cotter, R.; Golubiewski, A.; Showalter, A.M.; McCormick, S. and Bedinger, P. (2001). A leucine-rich repeat region is conserved in pollen extensin-like (Pex) proteins in monocots and dicots. *Plant Molecular Biology*, Vol.46, pp.43-56

Tang, B.; Banerjee, B.; Greenberger, P.A.; Fink, J.N.; Kelly, K.J. and Kurup, V.P. (2000). Antibody binding of deletion mutants of Asp f 2, the major Aspergillus fumigatus allergen. *Biochemical and Biophysical Research Communications*, Vol.270, pp.1128-1135

Thompson, J.D.; Higgins, D.G. and Gibson, T.J. (1994). CLUSTAL W: improving the sensitivity of progressive multiple sequence alignment through sequence weighting, position-specific gap penalties and weight matrix choice. *Nucleic Acids Research*, Vol.22, pp.4673-4680

Twell, D.; Wing, R.; Yamaguchi, J. and McCormick, S. (1989). Isolation and expression of an anther-specific gene from tomato. *Molecular and General Genetics*, Vol.217, pp.240-245

van Gunsteren, W.F.; Billeter, S.R.; Eising, A.A.; Hünenberger, P.H.; Krüger, P.; Mark, A.E.; Scott, W.R.P. and Tironi, I.G. (1996). Biomolecular Simulations: The GROMOS96 Manual and User Guide. Zürich, VdF Hochschulverlag ETHZ

van Ree R. (2004). Clinical importance of cross-reactivity in food allergy. *Current Opinion in Allergy & Clinical Immunology*. Vol.4, pp.235-240

Villalba, M.; Batanero, E.; Lopez-Otin, C.; Sanchez, L.M.; Monsalve, R.I.; Gonzalez de la Pena, M.A.; Lahoz, C. and Rodriguez, R. (1993). The amino acid sequence of Ole e I, the major allergen from olive tree (Olea europaea) pollen. *European Journal of Biochemistry*, Vol.216, pp.863-869

Villalba, M.; López-Otín, C.; Martín-Orozco, E.; Monsalve, R.I.; Palomino, P.; Lahoz, C. and Rodríguez, R. (1990). Isolation of three allergenic fractions of the major allergen from Olea europaea pollen and N-terminal amino acid sequence. *Biochemical and Biophysical Research Communications*, Vol.172, pp.523-528

Wagner, S. and Breiteneder, H. (2002). The latex-fruit syndrome. *Biochemical Society Transactions*, Vol.6, pp.935-940

The Information Systems for DNA Barcode Data

Di Liu and Juncai Ma
*Network Information Center, Institute of Microbiology, Chinese Academy of Sciences WFCC-MIRCEN World Data Centre for Microorganisms (WDCM)
China, People's Republic*

1. Introduction

DNA barcoding is a novel concept for the taxonomic identification, in that it uses a specific short genetic marker in an organism's DNA to discriminate species. In 2003, professor Paul D. N. Hebert, "the father of DNA barcoding", of the University of Guelph, Ontario, Canada first proposed the idea to identify biological species using DNA barcode, where the mitochondrial gene cytochrome *c* oxidase subunit I (COI) was supposed to be the first candidate for animals (Hebert et al. 2003a). Their studies of COI profiling in both higher taxonomic categories and species-level assignment demonstrated that COI gene has significant resolutions across the animal kingdom except the phylum Cnidaria (Hebert et al. 2003b, Ward et al. 2005, Hajibabaei et al. 2006). From then on, a wide broad of taxonomic groups (i.e. birds, fish, butterflies, spiders, ants, etc) were examined by COI gene for its usability as the barcode (i.e. Hebert et al. 2004a, Hebert et al. 2004b, Greenstone et al. 2005, Smith et al. 2005, Barber and Boyce 2006, Meier et al. 2006, Kerr et al. 2007, Kumar et al. 2007, Pfenninger et al. 2007, Stahls and Savolainen 2008, Zhou et al. 2009). Meanwhile, other candidate genes, including Internal Transcribed Spacer (ITS), trnH-psbA intergenic spacer (trnH-psbA), Ribulose-bisphosphate carboxylase (rbcL) and Maturase K (matK) were analysed by different research groups (Jaklitsch et al. 2006, Evans et al. 2007, Ran et al. 2010, de Groot et al. 2011, Liu et al. 2011, Piredda et al. 2011, Yesson et al. 2011). Till recently, there are about 30 DNA barcode candidates are tested, and 4 to 8 of them are widely used for the identification of diversified taxonomic groups with a relatively good resolution.

It has been estimated that there are 10 to 100 million species of living creatures in the earth, while what we know is very limited. Knowing the biodiversity is one of the crucial biological issues of ecology, evolutionary biology, bio-security, agro-biotechnology, bio-resources and many other areas. For very long period, taxonomists have provided a nomenclatural hierarchy and key prerequisites for the society. However, the needs for species identification requested by non-taxonomists require the knowledge held by taxonomists. Therefore, a standardized, rapid and inexpensive species identification approach is needed to establish for the non-specialists. There had some attempts on the molecular identification systems based on polymerase chain reaction (PCR), especially in bacterial studies (Woese 1996, Zhou et al. 1997, Maiden et al. 1998, Wirth et al. 2006), but no successful solutions for broader scopes of eukaryotes (reviewed in Frezal and Leblois 2008). The DNA Barcode of Life project is another attempt to create a universal eukaryotic identification system based on molecular approaches. Following studies by Hebert et al.

(Hebert et al. 2003a, Hebert et al. 2003b), the Consortium for the Barcode of Life (CBOL) was initiated in 2004, and aimed to produce a DNA barcode reference library and diagnostic tools based on the taxonomic knowledge to serve taxonomists and non-taxonomists (Schindel and Miller 2005). It should note that the DNA Barcode of Life project is neither to build the tree of life nor molecular taxonomy (Ebach and Holdrege 2005, Gregory 2005). From the establishment of CBOL, more than 50 countries have been participated in and devoted themselves into this ultimate mission. One of the important projects is the International Barcode of Life project (iBOL) sponsored by Canada government (details will be described below). Till now, DNA barcoding is accepted by a great range of scientists and has achieved indisputable success (Teletchea 2010).

One of the major aims of bioinformatics is to finely store and manage the huge amount of biological data. Apart from genome sequencing projects, DNA barcoding projects are going to establish another important biological data resource to the public. Until now, there are about half a million DNA barcodes are submitted to GenBank from the Barcode of Life Data System (BOLD) (Ratnasingham and Hebert 2007), the most essential data center for barcode of life projects. Besides, large amount of DNA barcode data are under producing and to be released worldwide. It has been estimated that more than 100 million barcode records will be generated for the animal kingdom(Ratnasingham and Hebert 2007), and that size is comparable to the current GenBank release (Benson et al. 2011). Unlike the traditional nucleotide sequences deposit in the international nucleotide sequence databases collaboration (INSDC), DNA barcode data comprises comprehensive data types, including photos, DNA chromatogram (trace files), geographic data and structured morphological information of each specimen. Therefore novel information systems are required to be developed to collect, store, manage, visualize, distribute, and utilize these data for species identification, clustering/classification as well as evolutionary studies. Moreover, applying the second-generation sequencing technology (e.g. Roche 454) for DNA barcoding, especially for those environmental samples (e.g. mud, water) is under developing, and this will generate a large amount of DNA barcodes a time, with the data files different from those from traditional DNA analyser implementing Sanger sequencing approach. Hence, methods to manage and utilize the output from 2nd-generation sequencers are also to be developed. Besides, it is still a great challenge to integrate the DNA barcode data into the studies of metagenomics (Venter et al. 2004, Rusch et al. 2007).

In this chapter, we will first review the current progresses of DNA barcode of life projects, and then we will describe the data schema and the information systems of DNA barcode data. Particularly, three types of DNA barcode information systems are to be introduced: BOLD, by now the best information system for DNA barcoding with highly integration; Global Mirror System of DNA Barcode Data (GMS-DBD), the mirror system for the distribution of the publicly available DNA barcode data worldwide; and the management system for Chinese DNA barcode data, which is a manageable information system for DNA barcoding groups.

2. DNA barcode and the international collaborations

In order to make the concept that using DNA barcodes to identify species being reality, great efforts need to be contributed by the nations. After the launch in 2004, CBOL has gathered more than 150 organizations around the world, including natural history museums, zoos, herbaria, botanical gardens, university departments, governmental organizations and private companies. The goals of CBOL include building up a DNA barcode library of the eukaryotic lives in 20 years. In July 2009, the Ontario Genomics Institute (OGI), Canada initiated the iBOL project, which was considered as the extension and expansion of the previous project, Canadian Barcode of Life Network (http://www.bolnet.ca) launched in 2005. Nowadays, there are 27 countries as partner nations participated in this international collaboration and nearly 20 established campaigns were co-working for iBOL on some specific creatures.

2.1 The concept of DNA barcode and the commonly used ones

DNA barcode is a segment of DNA that possesses the following features. a) DNA barcode is conserved in a broad range of species, so that a conserved pair (or several conserved pairs) of primers can be designed and applied for DNA amplification; b) it is orthologous; c) DNA barcode must evolve rapidly enough to represent the differences between species; d) DNA barcode needs to be short, so that a single DNA sequencing reaction is enough to obtain the sequence; and e) DNA barcode needs to be long to be capable of holding all substitutions within higher taxonomic groups (For example, a 500-base pair (bp)-long DNA barcode has the capability to hold 4^{500} possible differences to discriminate species.).

The first DNA barcode is a 658-bp long region DNA segment of COI gene within mitochondria, with primers LCO1490 (5'-GGTCAACAAATCATAAAGATATTGG-3') and HCO2198 (5'-TAAACTTCAGGGTGACCAAAAAATCA-3') used for DNA amplification (Hebert et al. 2003a). COI gene as the primary DNA barcode has been proven to be useful in broad ranges of animal species, despite of the limitations in some taxa (Meyer and Paulay 2005, Vences et al. 2005). In fungi, ITS was chosen as the main DNA barcode and was confirmed by the sequences within the international nucleotide sequence databases (Nilsson et al. 2006), though COI was examined applicable in Penicillium (Seifert et al. 2007). In plants, mitochondrial DNA shows intra-molecular recombination and COI gene has lower evolutionary rate (Mower et al. 2007), so that genes on the plasmid genome were examined, e.g. *rpoB*, *rpoC1*, *rbcL* and *matK*. Meanwhile, some intergenic spacers (e.g. trnH-psbA, atpF-atpH and psbK-psbI (Fazekas et al. 2008)), and markers' recombination (e.g. *rbcL* and trnH-psbA (Kress and Erickson 2007)) were tested, too. Nevertheless, those choices either meet the amplification problems or standardization problems. Recently, CBOL Plant Working Group recommended the combination of *rbcL* and *matK* for plant DNA barcoding (CBOL Plant Working Group, 2009).

2.2 The international Barcode of Life project

The main mission of iBOL is "extending the geographic and taxonomic coverage of the barcode reference library -- Barcode of Life Data Systems (BOLD) -- storing the resulting barcode records, providing community access to the knowledge they represent and creating new devices to ensure global access to this information." (http://ibol.org/about-us/what-is-ibol/). To accomplish the mission step by step, iBOL announced the first 5-year plan that is to collect and process 5 million samples covering 500 thousand species with $150 million budget. Then six working groups were established to work on barcode library construction, methodology, informatics, technology, administration and social issues (Table 1). The first two working groups are mainly focusing on the collection and production of DNA barcodes in various living creatures, biotas and specimens in museums. The third working group is dedicated to the construction of the informatics, including the core functionality and mirror sites. Core functionality comprises at least a sophisticated bioinformatics platform with the integration of a robust IT infrastructure (computational note, storage and network), DNA barcode databases and analytical tools. Meanwhile, the mirror sites help to strengthen data security and accessibility. Working group 4 is focusing on future technologies, either applying the latest sequence techniques or developing the portable mobile devices. Although working groups 5 and 6 are not purely on the science and technology of DNA barcoding, the administration and dealing with social aspects are far more important to the success of the project.

iBOL Working Group	Sub-Working Group
WG 1. Barcode Library: Building the digital library of life on Earth	WG 1.1, Vertebrates
	WG 1.2, Land plants
	WG 1.3, Fungi
	WG 1.4, Animal Parasites, Pathogens & Vectors
	WG 1.5, Agriculatural and Forestry Pests and Their Parasitoids
	WG 1.6, Pollinators
	WG 1.7, Freshwater Bio-surveillance
	WG 1.8, Marine Bio-surveillance
	WG 1.9, Terrestrial Bio-surveillance
	WG 1.10, Polar Life
WG 2. Methods: Extending the horizons of barcoding	WG 2.1, Barcoding Biotas
	WG 2.2, Museum Life
	WG 2.3, Methodological Innovation
	WG 2.4, Paleobarcoding
WG 3. Informatics: Storing and analyzing barcode data	WG 3.1, Core Functionality
	WG 3.2, Mirrors
WG 4. Technologies	WG 4.1, Environmental Barcoding
	WG 4.2, Mobile Barcoding
WG 5. Administration: Consolidating the matrix	WG 5.1, Project Management
	WG 5.2, Communications
WG 6. GE³LS	WG 6.1, Equitable Use of Genetic Resources
	WG 6.2, Regulation and International Trade
	WG 6.3, Intellectual Property and Knowledge Management
	WG 6.4, Education Initiatives for Schools and Media
	WG 6.5, Governance of Knowledge Mobilization

Table 1. iBOL working groups.

The Information Systems for DNA Barcode Data 183

By the end of the year 2010, iBOL reported the progresses of each working group in the iBOL Project Interim Review (http://ibol.org/interim-review/). During the first 18 months, iBOL has produced DNA barcodes for 153K species from 326K specimens collected worldwide, and obtained exciting results from barcoding biotas of the locales Moorea and Churchill, where a great number of additional species were revealed by DNA barcoding. BOLD as the core functionality of iBOL has increased the number of records to 1.1 million and the number of users to 6000. The power of storage and computing was also improved dramatically. Conclusively, all working groups have made substantial progresses towards the final goals.

The achievements made by iBOL are with the help of the campaigns of barcode of life, which consists of researchers with similar interests on specific families and regions of life (e.g. birds, fish, etc.). Most of the campaigns are working closely with the relevant iBOL working groups and/or BOLD. Below lists some useful websites and campaigns of the international collaborations (Table 2).

Short Name	Description	URL
CBOL	The consortium for the barcode of life	http://www.barcoding.si.edu; http://www.barcodeoflife.org
iBOL	The international barcode of life project	http://www.ibol.org
CCDB	Canadian centre for DNA barcoding	http://www.danbarcoding.ca
BOLD	Barcode of life data systems	http://www.boldsystems.org
GMS-DBD	Global mirror system of DNA barcode data	http://www.boldmirror.net
Fish-BOL	Fish barcode of life initiative	http://www.fishbol.org
ABBI	All birds barcoding initiative	http://www.barcodingbirds.org
PolarBOL	Polar barcode of life	http://www.polarbarcoding.org
Bee-BOL	Bee barcode of life initiative	http://www.bee-bol.org
MarBOL	Marine barcode of life	http://www.marinebarcoding.org
	Lepidoptera barcode of life	http://lepbarcoding.org
	Trichoptera barcode of life	http://trichopterabol.org
	Formicidae barcode of life	http://www.formicidaebol.org
	Coral beef barcode of life	http://www.reefbarcoding.org
	Mammal barcode of life	http://www.mammaliabol.org
	Sponge barcoding project	http://www.spongebarcoding.org

Table 2. Websites of DNA barcode of life projects worldwide

2.3 DNA barcode of life projects in China

There are three categories of the participated nations of iBOL, the National Nodes, the Regional Nodes and the Central Nodes. The National Node is primarily to collect, identify and curate the specimens from their territory, and the Regional Node has additional duties to participate in DNA barcode acquisition. As for a Central Node, it has not only National Node and Regional Node's missions, but also to maintain core sequencing facilities and the bioinformatics facilities, as well as to help share DNA barcode records with all nations. Of the current 27 nations participated in iBOL (27 nations are shown in iBOL website, where there are 33 nations in the iBOL Project Interim Review), China is acting as one of the four

Central Nodes, while the others are Canada, United States, and the European Union (France, Germany, Netherlands, etc.).

To better support the international collaborations and to take great part in the iBOL project, China has established the China National Committee for iBOL project. Prof. Jiangyang Li, Vice President of Chinese Academy of Sciences (CAS), acts as President, and Prof. Zhibin Zhang of the Bureau of Life Sciences and Biotechnology, CAS and Prof. Yaping Zhang of the Kunming Institute of Zoology, CAS are taking the roles of Vice President. From then on, the constructions of the core sequencing facilities and bioinformatics facilities for DNA barcoding were initiated, and varied foundations of China, including CAS, the Ministry of Science and Technology (MOST) and the Natural Science Foundations of China (NSFC) have issued projects of DNA barcode of life (Table 3). The projects covered the studies on diversified creatures, including animal, plant and fungus, and the studies on specimen collection, data production, theory and methodology, database and information system, etc. To date, China has established three main research campaigns, working on animal, plant and fungal DNA barcoding respectively, and initiated the constructions of China DNA Barcoding Centre and China DNA Barcode of Life Information Centre by the institutes of CAS.

Projects issued by	Projects aims at
Chinese Academy of Sciences (CAS)	Specimen collection; DNA barcode data production; Basic research of DNA barcoding; Construction of information centre and central database of China National Committee for iBOL Project.
Ministry of Science and Technology (MOST)	Studies on animal, plant and fungal barcoding; Construction of DNA barcode databases and information systems
Natural Science Foundations of China (NSFC)	Basic research on DNA barcoding theories and methodologies for animal, plant and fungal barcoding

Table 3. DNA barcode of life projects in China

In terms of the construction of China DNA Barcoding Information Centre, we are now running projects from CAS and responsible for the initiation and implementation. With the help from the research campaigns of China and iBOL, we have designed the architecture of the Chinese DNA Barcode of Life Information Systems. Briefly, the entire systems consist of two essential components, the Mirror System of BOLD and the Management System for China DNA barcoding projects (Fig. 1). Each component is an independent system based on the data it contains, and serves as separated services. Nevertheless, the China DNA Barcode of Life Information Centre maintains a DNA barcode database that integrates the data of BOLD mirror and the Chinese DNA barcode data. In views of functions, the mirror system will mainly focus on data synchronization, data presentation, and statistical and analytical tools, while the data management system focuses on data submission, data verification, and data publishing. In sections below, we will describe these two kinds of systems in details.

The Information Systems for DNA Barcode Data 185

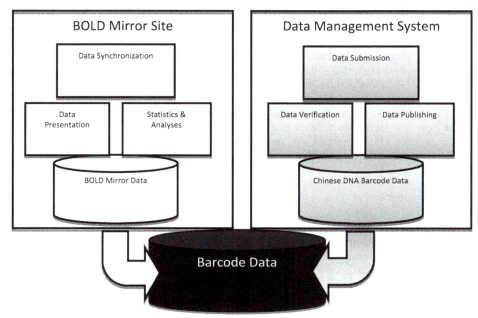

Fig. 1. The architecture of the Chinese DNA Barcode of Life Information Systems. The left square represents the functions of the Mirror System of BOLD, and the right square shows the functions of Data Management System. The DNA barcode data of both systems are further integrated into a centralized DNA barcoding database.

Fig. 2. Homepage of Barcode of Life Data System (BOLD).

3. DNA barcode data schema and the Barcode of Life Data systems

The Barcode of Life Data System (BOLD) is by far the best information system for DNA barcode data, aiming at the collection, management, analysis and use of DNA barcodes (Ratnasingham and Hebert 2007). Started from 2004, BOLD has developed to not only the authorized DNA barcode data resource, but also a global workbench for the assembly, analysis and publication of DNA barcode records. By now, BOLD has become a major contributor to INSDC, with the contribution of about 500,000 records into GenBank during 2010. Millions of DNA barcode records will be deposited into INSDC along with the proceedings of iBOL project. Each record submitted into GenBank has the keyword "BARCODE" and an ID back to BOLD.

Since BOLD uses different ways to store and present DNA barcoding data to GenBank, in the following sections we will give a brief dissection of BOLD, from data schema to the functions. To some extent, this will also help to understand another information systems to be introduced in this chapter.

3.1 Data schema of barcode of life data

In order to collect and manage the DNA barcode data effectively, a data schema is required. BOLD has built up its data schema according the Darwin Core 1 standard, which is applied by the Global Biodiversity Information Facility (GBIF) and other biodiversity alliances, for the data fields related to specimen description. Meanwhile, the data schema describes the format for the sequence (barcode or marker) information as well as the primers and trace files. In brief, there are three categories of information, specimen related, sequence related and primer related (summarized in Table 4, and example in Fig. 3). Specimen related information includes the voucher info, collection info, taxonomic info and details, as well as some multi-media files (mostly photos in the current stage). Sequence related information consists of sequence file (in FASTA format), primer codes and trace files. Considered that primer pairs to the markers (DNA barcodes) are to be standardized and the dataset is relatively small and constant, the detailed info of the primer is separated from specimen and sequence.

As far as a DNA barcode record is concerned, the *Sample ID* is one of the most important key fields. It is the identifier of the sample of specimen used for sequencing, so that it is unique in the whole system. Different *Sample IDs* may refer to one specimen. For example, two legs of butterfly are treated separated for experiments. Since there is not a single field like "Specimen ID" to uniquely mark specimens, BOLD schema uses *Field ID*, *Museum ID* and *Collection Code* instead. At least one of the *Field ID* and *Museum ID* must be appeared and must associate with *Institution Storing* to exclusively locate the specimen in the world. A *Collection Code* is required whenever it has to be combined with *Museum ID* to discriminate specimen. That is the basis to link a DNA barcode to a real specimen. The taxonomic info is for the link between DNA barcode and taxon assigned. Considered that some samples are difficult to identify (e.g. damaged organisms, immature specimen), a full taxonomic assignment is not mandatory, but the phylum level assignment is a prerequisite. Collection info is essential for knowing the global distribution of a specific taxon and the variations among different areas, so that detailed geographical information is encouraged to provide. Additionally, the Details describe the specimen in detail and the Images show the morphological natures. Another key field is *Process ID*, which is used to identify the experimental process that produces a DNA barcode. One *Sample ID* is uniquely referred to one *Process ID*, vice versa. This ensures the connection between sample and produced DNA. *Process ID* is also known as *Barcode ID* in the sequence record view. Finally, trace files are essential to qualify the results of DNA sequencing, and primer info is required when a process needs to be repeated.

1st Category	2nd category	Data fields	Description
Specimen	Voucher info	Sample ID	ID associated with the sample being sequenced. It is unique.
		Field ID	Field number from a collection event or specimen identifier from a private collection.
		Museum ID	Catalog number in curated collection for a vouchered specimen.
		Collection Code	Code associated with given collection.
		Institution Storing	Full name of the institution where specimen is vouchered.
		Sample Donor	Full name of individual responsible for providing specimen or tissue sample.
		Donor Email	E-mail of the sample donor.
	Taxonomic info	Taxonomy	Full taxonomy. Phylum is mandatory.
		Identifier	Primary individual responsible for the taxonomic identification
		Identifier Email	Email address of the primary identifier
		Identifier Institution	Institution of the primary identifier
	Collection Info	Collectors	List of collectors
		Collection Date	Data of collection
		Continent/Ocean	Continent or ocean name
		Country	Country name
		State/Province	State and/or province
		Region & Sector & Exact site	Detailed description of place
		GPS	GPS coordinates
		Elevation/Depth	Elevation or depth
	Details	Sex	Male/female/hermaphrodite
		Reproduction	Sexual/asexual/cyclic pathogen
		Life Stage	Adult/immature
		Extra Info & Notes	User specified, free text
	Images	Image File	Name of image
		Original Specimen	If the image is from the original specimen

1st Category	2nd category	Data fields	Description
		View Metadata	Dorsal/Lateral/Ventral/Frontal/etc.
		Caption	Short description of image
		Measurement	Measurement that was taken
		Measurement Type	Body length, wing span, etc.
		Copyright	The copyright
Sequence	Sequence	Process ID (Barcode ID)	The ID of a process that produce a sequence
		Sequence	DNA sequence
	Trace Files	Trace File	Complete name of trace file
		Score File	Complete name of score file
		Read Direction	Forward or reverse
		Marker	COI-5P, ITS, rbcLa, matK, etc.
		PCR Primer Codes	PCR primers used
		Sequence Primer Codes	Sequence primers used
Primers		Primer Code	Unique code for a primer
		Primer Description	A description of what the primer is used for
		Alias Codes	Any other known codes
		Target Marker	COI-5P, ITS, etc.
		Cocktail Primer	If it is a cocktail primer
		Primer Sequence	Sequences
		Direction	The direction of the sequence
		Reference/Citation	References and/or citations
		Notes	Some notes

Table 4. Summary of the main data fields in BOLD data schema.

3.2 Barcode of Life Data system

BOLD system consists of three main modules, the Management and Analysis System (MAS), Identification System (IDS)/identification engine, and External Connectivity System (ECS). MAS is responsible for data repository, data management, data uploads, downloads and searches and some integrated analytics (Ratnasingham and Hebert 2007). With no doubt, it comprises the most important functions. According to the data schema described above, it stores specimen related information, sequences, trace files and images. All data of records was uploaded and organized by project that is created by the user. Once a user creates a project for a set of DNA barcode records, at least two data fields, *Sample ID* and *Phylum*, for each record are to be filled. Then additional information including voucher data, collection info, taxonomic assignment, identifier of the specimen, >500-bp sequence of DNA barcode,

PCR primers, and trace files is needed for a full data record. Among all DNA barcode records stored in BOLD, not all are complete and in high quality. For example, there are sequences with more than 1% Ns or less then 500-bp. Hence the integrated analytic tools are useful to help find out those records with low quality. In brief, BOLD employs Hidden Markov Model (Eddy 1998) (on amino acids) to align sequences and then to verify if the correct gene sequence was uploaded; consequently, scripts are used to check for stop codon and to compare against possible contaminant sequence. For trace file, a mean Phred score (Ewing and Green 1998) for the full sequence is determined, and this is used for the quality categorization. After these processing, a record will be flagged if has missing fields, sequence error or low quality.

```xml
<record>
  <recordID>1224108</recordID>
  <processid>ASANR583-09</processid>
  <specimen_identifiers>
    <sampleid>CASENT0042697-D01</sampleid>
    <catalognum>CASENT0042697-D01</catalognum>
    <fieldnum>BLF09080</fieldnum>
    <institution_storing>California Academy of Sciences</institution_storing>
  </specimen_identifiers>
  <taxonomy>
    <identification_provided_by>Brian Fisher</identification_provided_by>
    <phylum>
      <taxon>
        <taxID>20</taxID>
        <name>Arthropoda</name>
      </taxon>
    </phylum>
    <class>
      <taxon>
        <taxID>82</taxID>
        <name>Insecta</name>
      </taxon>
    </class>
    <order>
      <taxon>
        <taxID>125</taxID>
        <name>Hymenoptera</name>
      </taxon>
    </order>
  </taxonomy>
  <collection_event>
    <collectors>B.L.Fisher</collectors>
    <collectiondate>2003-11-18</collectiondate>
    <coordinates>
      <lat>-14.443</lat>
      <lon>49.743</lon>
      <coordsource></coordsource>
      <accuracy></accuracy>
    </coordinates>
    <region>Malagasy</region>
    <exactsite>Parc National de Marojejy, Antranohofa, 26.6 km 31deg NNE Andapa, 10.7 km 318deg NW Manantenina</exactsite>
    <country>Madagascar</country>
    <province>Antsiranana</province>
  </collection_event>
  <sequences>
    <sequence>
      <sequenceID>3378849</sequenceID>
      <markercode>COI-5P</markercode>
      <genbank_accession>GU711306</genbank_accession>
      <nucleotides>
------ATTCACTAATTAATAATGACCAAATTTATAACTCTCTAATTACTAGGCACGCCTTAATTATAATTTTTTTTATAATTATACCTTTTA
TAATTGGAGGATTTGGAAATTTCCTAGTCCCACTAATACTAGGGGCCCCTGATATAGCCTACCCTCGTATAAATAACATAAGATTCTGACTAT
TGCCCCCTTCCCTAATACTTTTAATTAGAGGAAGATTTATTAGAGATGGAGTAGGAACAGGATGAACCATCTATCCCCCCCTTTCATCAAATA
TTTTCCATAACGGCCCTTCTGTAGACCTTTCAATTTTCTCACTTCATATCGCAGGAATATCTTCTATTTTAGGAGCTATTAACTTTATTTCAA
CTATTATTAATATAAAAAACTCTGGCCTATCATTAGACAAAATTTCATTACTAATCTGATCAATCAACATCACCGCTATTCTCTTACTTCTCT
CCTTACCAGTCTTAGCCGGAGCAATTACTATATTATTTACGGATCGTAATTTAAACACTTCTTTTTTTGACCCATCAGGAGGGGGAGATCCTA
TTTTATTCCAACATTTATTT</nucleotides>
      <last_updated>2011-04-10T13:04:27Z</last_updated>
    </sequence>
  </sequences>
  <last_updated>2011-04-10T13:04:27Z</last_updated>
  <notes></notes>
</record>
```

Fig. 3. An example of an XML format DNA barcode records according to BOLD data schema.

The IDS is one of the most commonly used analytic tools of BOLD. It uses all sequences uploaded, both public and private ones, to locate the closest match. Note that the details of the private ones are not exposed. As for animal identification, COI gene set is used as database to compare against. The Basic Local Alignment Search Tool (BLAST) (Altschul et al. 1990) is employed to detect single base indels, and Hidden Markov Model for COI protein is used for sequence alignment. There are four databases are used for COI identification in BOLD, including All Barcode Records Database, Species Level Barcode Database, Public Record Barcode Database and Full Length Record Database. They are comprised of different quality levels of sequences (http://boldsystems.org/docs/handbook.php?page=idengine). Fungal identification is based on ITS, and plant identification is on *rbcL* and *matK*. The Fungal Database and Plant Database respectively are for the identification and only BLAST algorithm is employed. By now, the data records within fungal and plant databases are much fewer than those in COI databases.

Besides IDS, there are other useful tools developed and integrated in BOLD. The Barcode Index Number system (BINs) is designed as an alternate method for species identification. In BINs, a set of operational taxonomic units (OTUs; putative species) was generated using a novel clustering algorithm based on graph methods, and a BIN ID is given for each OTU. BINs and OTUs help to solve the problem that many BOLD records have only interim species name or without fully taxonomic assignment. Another tool, the Taxon ID Tree employs varied distance metrics to build neighbour-joining tree with at most 5000 species a time. This is powerful toolbox for online phylogenetics analysis. More functions including distance summary, sequence composition, nearest neighbour summary, DNA degradation test, accumulation curve and alignment viewer are available in BOLD. These tools implemented the bioinformatics and statistic approaches for data analyses.

ECS is served as the interface for other developers to access the barcode data via web services. Currently, BOLD opens two services e-Search and e-Fetch following the Representational State Transfer (REST) architecture. Programmes may use e-Search to get a list of records, and use e-Fetch to obtain the details. Below lists the parameters for e-Search and e-Fetch. Anther web service called eTrace is also developed for the retrieval of trace files for a given Sample ID. We have tested for the use of this service, and obtained thousands of trace files from BOLD. This service will be exposed to public in the near future.

Service Name	Parameters	Description
e-Search & e-Fetch	id_type	Sample_id, process_id, specimen_id, sequence_id, tax_id, record_id
	ids	Comma separated ids
	Return_type	Text, xml, json
	File_type	Zip
e-Search	Geo_inc	Country/province to be included
	Geo_exc	Country/province to be excluded
	Taxon_inc	Taxonomy to be included
	Taxon_exc	Taxonomy to be excluded
e-Fetch	Record_type	Specimen, sequence, full

Table 5. Parameters of the web services of BOLD.

4. The Global Mirror System of DNA Barcode Data (GMS-DBD)

One of the main tasks of iBOL project is to setup global mirror sites for DNA barcode data, and this is assigned as the mission of working group 3.2 (that is chaired by the author Juncai Ma). Mirror sites play roles not only for data security but also for the global access and use of DNA barcode data fast and stably. In addition to iBOL's task, the Chinese DNA Barcode of Life Information Systems requires to mirror BOLD as well. For the current stage, the entire BOLD system is difficult to mirror, in both the storage and the analytical workbench. For this reason, we developed the mirror site of BOLD data (http://www.boldmirror.net) (Fig. 4) in China in 2010, and served it as one of the major components of the Chinese DNA Barcode of Life Information Systems. In late 2010, we started to encapsulate the mirror site into a distributable mirror system, namely the Global Mirror System of DNA Barcode Data (GMS-DBD). Different from BOLD systems, GMS-DBD is designed and currently served as a system for the presentation and analysis of DNA barcode data only, but not the management of DNA barcode projects. Moreover, GMS-DBD is designed to feature in the fast deployment and fast use of the DNA barcode data.

Fig. 4. Homepage of the mirror of BOLD data.

4.1 Design and implementation of GMS-DBD

For the purposes mentioned above, GMS-DBD was designed with three main components including data synchronization module, data presentation module, and statistics and analysis module. As for data synchronization, the module functions to obtain data from BOLD and then to distribute them to each mirror site deployed by GMS-DBD. Right after data transferred over Internet, all updated data are imported automatically into mirror site's local database management system (DBMS). Data presentation module comprises the following functions, browsing by records, browsing by taxonomy and searching by keywords. These are the key functions for a DNA barcode data mirror. Data statistics module aims at the statistical presentation of the entire dataset, including the barcode data statistics by country, by taxon or by organization. (Fig. 5 Left) For the use of DNA barcode data, especially for sequence similarity based identification, we embedded a form to submit the sequences to be identified to a BLAST server in China mirror site.

The whole system was implemented as a software package, in which the database, applications, web pages were encapsulated. This software package was for the Linux platform, and tested on some main distributes, like Fedora, Ubuntu. The installer was written in Perl, and will guide the administrator to setup and configure the mirror site step-by-step. Particularly, Apache web server and PHP scripting language were employed for the Web layer, and MySQL DBMS was used for the management of database. The applications for database search, records presenting and sorting, and statistics calculation were written using PHP scripts. For better visualization, Java applet (viewing chromatogram files), Adobe FLEX (statistical presentation) and Google Maps (geographical view of locations) were employed and embedded into the web pages. In addition to presenting record by specimen information and by sequence information as BOLD does, we developed a full record view to browse all information in a single page (Fig. 5 right).

In additional to the installer, data presentation module, and statistics and analysis module, data synchronization module was developed separately. From late 2009, BOLD and our centre were beginning to test the transfer of DNA barcode data between Canada and China, and finally defined a method. Every day, BOLD dumps the daily-updated data and daily-completed data into XML format files and put them on a HTTP server, and for the mirror site, we run a daemon process to download the latest files. After data transfer finished, the daemon process will invoke consequently another processes to parse the XML format files and to import the parsed files into MySQL database. Perl is used for the parser and Structured Query Language (SQL) is used for data import. Specific for the data import process, it will execute the insertion and updating of the new entries and log every modification of the whole dataset. In parallel, another process will run to extract the DNA sequences of the new records and then to index them for BLAST search. All these procedures are scheduled and run automatically, and the updates and logs will be shown on the web pages immediately after the procedures finished.

4.2 Distribution of GMS-DBD and DNA Barcode data

GMS-DBD is freely distributed as the DNA barcode data. Nowadays, the University of Waikato, New Zealand has firstly built up their mirror site (http://nz.boldmirror.net) using the GMS-DBD distributes, and gave us great suggestions on the improvement and the further development of GMS-DBD. The Centro de Excelencia em Bioinformatica (CEBio) of Brazil has also contacted us and is setting up the mirror site using GMS-DBD.

Fig. 5. Functions of GMS-DBD. Top-left, logs of data updating. Middle-left, view by taxonomic tree. Bottom-left, statistics by phylum. Right, full DNA barcode record.

To date, there are ~636,000 DNA barcode records available for the mirror sites, though more than 1.2 million records deposit in BOLD. One reason is that mirror sites stored the public available data records only, while BOLD has some private data to be released in the future. The second is that there are incomplete records held in BOLD that did not distribute to the mirror sites. Among all the records within the mirror sites, more than 200 thousand of them have trace files. The photos for each specimen are not available for mirror sites by now, because there are copyright problems to be solved.

5. Design and implementation of the DNA barcode management system in China

As described in the previous sections, China has launched several projects to contribute the construction of the global DNA barcode library. An information system is thus needed to collect, manage and publish the produced data. Although BOLD is a good choice for the management of users' DNA barcode projects and data records, many scientists are still willing to hold their data before their results published in scientific journals. Therefore, we designed and developed a DNA barcode management system for the Chinese scientists to manage their barcode data. First, the system is also designed as a distributable version, and could be downloaded and installed locally. Second, the user can hold the data for privacy for long time. Third, it supports modifications for the data schema. Fourth, for its simplicity, it lacks the connectivity to any LIMS, but using a unified data format to exchange data from LIMS.

In views of function, this system has similar aspects to the BOLD system, i.e. user identification system, project based management, data input forms for different data types, data review and analyses platform. The entire system is developed with the LAMP (Linux+Apache+MySQL+PHP) architecture, and of no need to mount on very heavy computing infrastructures. The system manager has the privileges to choose the hardware ands scale the capability of the system. For data's safety, only the registered users are allowed to use. The registered users have their data records organized by project, and only their own projects or authorized ones are allowed to visit (Fig. 6 left). The data input forms are like BOLD's as well, in that there are forms for specimen info, taxonomy info, sequences, photo, trace files, etc., except that all the labels have Chinese translations.

Fig. 6. Snapshots of DNA barcode management system in China. Left, user's project page, summarizing the projects and records. Right, Quick Upload Page.

BOLD has exemplified the management of DNA barcode data as a centralized data centre and shall be the reference for the development of core functionality of the Central Nodes. However, the development of such a comprehensive system needs a long period, and inapplicable for the immediate use. Moreover, things are to some extent different in China than in Canada. First, there is no constructed barcoding centre like CCDB before varied DNA barcoding projects were issued. Second, the research campaigns studying animal, plant and fungus barcoding respectively have already defined the working structure for each studies, especially for the specimen and collection information. Additionally, the DNA sequences produced by each campaign are either stored in their LIMS or simply stored in personal computers. In this situation, the tough work is becoming how to make the DNA barcode data management system suitable for every data structure, and how to easily collect those data already there. To meet this need, we designed different data schema for animal, plant and fungus. Note that all data schema are following the latest BOLD data schema, but with some data fields modified to fit each species groups. For example, we omitted the data fields "sex", "reproduction" and "life stage" for fungi, but added on data fields "habitat" and "storing method". This was the result after discussion with scientists doing fungus research. Additionally, we also added on some fields for Chinese, like "Chinese common name", "taxon name in Chinese", etc. Moreover, we encouraged the users to use the Chinese characters for "provider name" and "location names", as they might be ambiguous in English.

Another feature of this data management system is that we implemented a gateway for quick upload (Fig. 6 right). In terms of the data already produced, they are stored and organized either by a local DBMS or by Excel datasheet with files. The providers would like to upload them into the system in batch mode, but not form by form. Then we developed the quick upload gateway, according to the data schema and created different templates for batch upload. In brief, the template is in Excel format, and every data field is in one column while every record is in one row. What the user needs to do first is to fill in the template for their data or slightly modify the datasheet in use. Then three files are required to prepare, one is zipped photos, another is zipped trace files, and the other is FASTA format sequence file. Note that the file names of photos and trace files, and the sequence name (*Process ID*) of the sequence should be in the right place of the Excel template. Then these four files may be uploaded onto the system and records are imported into the database in batch mode. This mode is also suitable for those who are in charge of the whole procedures of the production of DNA barcode data, from specimen to barcode.

Functions of the GMS-DBD were applied for this management system, in that we integrated Google Maps for geographical presentation, Java applet for viewing trace files, and BLAST for sequence identification. Besides, the management system has a data publishing function, which can generate XML format data following BOLD data schema. Within data transformation (from MySQL to XML), a language translation module will automatically invoked to translate those Chinese "city names", "institution names" and "person names" into English.

By now, this DNA barcode data management system is implemented using Chinese language, and has been developing for multi-language uses. To date, it is used for the collection and management of DNA barcodes of fungi, fish, birds, amphibian, and plants in China.

6. Perspectives of DNA barcode information system and the underneath bioinformatics

Along with the success of DNA barcode of life projects worldwide, huge amount of data will be produced on purpose. The traditional sequence database like GenBank seems not suitable for the storage of the DNA barcode records with multiple data types, so that novel systems are required to develop for the management and utilization of those data. Besides the information systems described above, DNA Data Bank of Japan (DDBJ) (Kaminuma et al. 2011) maintains the Japanese repository for barcode data and employed the its BLAST server for identification, and the Korean BioInformation Center of Korea Research Institute of Bioscience and Biotechnology (KRIBB), Korea developed the BioBarcode platform for Asian biodiversity resources (Lim et al. 2009). The CBS culture collection of Netherlands is attempting to integrate the DNA barcode data into the BioloMICS software, which bundles comprehensive bioinformatics tools for data analyses. This will provide better experiences in some aspects for the bioinformatics researches applying barcode data, though the online workbench of BOLD provides sets of approaches for data analysis.

When bioinformatics is mentioned, the algorithms and software are first recalled. The commonly used method for species discrimination is the Neighbour Joining (NJ) algorithm with Kimura 2 parameters (K2P) corrections. Though this approach was claimed as the best DNA substitution model for close genetic relations (Nei and Kuman 2000), the maximum likelihood methods and Bayesian Inference are more and more used for DNA barcoding analysis (e.g. Mueller 2006, deWaard et al. 2010, Kim et al. 2010). Coalescent-based methods for phylogenetics were also examined for DNA barcoding (Nielsen and Matz 2006, Abdo and Golding 2007). Casiraghi *et al.* (Casiraghi et al. 2010) summarized bioinformatics approaches for the analyses of barcode data and proposed the use of varied methods for different scenarios.

Although DNA barcoding technology has been largely improved and the DNA barcode data has rapidly accumulated, there are still some concerns on the use of DNA barcode. First, varied paired of primers might be used to identify an unknown sample or a mixture. Although COI gene was proved to be efficient in almost all animals and some groups of fungi, the identification of unknown specimen may need several paired of primers and this will increase the complexity of the automation of DNA barcoding process. Prospectively, this would be one of the major issues on the development and implementation of the handy device for DNA barcoding, as planned in iBOL WG4.2. Secondly, every DNA barcode has the limitation of resolution in specific species groups, so that auxiliary markers need to be discovered. This problem exists mainly in plant and fungi, though some animal groups have met same one. Currently, plant and fungi data deposited in BOLD are still limited, and a lot of groups need to be examined. Additionally, the approaches for sharing and using barcode data need to be improved. A user-friendlier interface to access the barcode dataset is needed. For instance, the well-examined DNA barcodes and/or the consensus of barcodes of each taxon are organized and prepared, and user needs only to select the interested ones and download only a small dataset. This will be convenient for the users (i.e. identification of specimen using DNA barcoding) and some researchers on barcoding. As a matter of fact, that is one of the tasks what BOLD and our group are working on.

7. Conclusions

With no doubt, DNA barcoding is becoming a popular approach for knowing the biodiversity on the earth, by utilizing the accumulative knowledge of taxonomy, the modern techniques of molecular biology and bioinformatics. Bioinformatics played prominent role in the construction and the employment of the global barcode of life library, from the management of data to the development of novel methods.

8. Acknowledgments

The authors thank the team members in the Network Information Center, Institute of Microbiology, Chinese Academy of Sciences. The work in our centre is supported, in part, by the projects from the Bureau of Life Sciences and Biotechnology of CAS, the Project of Informatization of CAS, and the Project of Scientific Databases of CAS (http://www.csdb.cn). Our centre is also obtained support from the State Key Laboratory of Microbial Resources (SKLMR), Institute of Microbiology, CAS. DL would also like to appreciate the support from Natural Science Foundations of China (NSFC) (Grant no. 30800640). The authors appreciate the supports and advices from iBOL, CCDB and BOLD.

9. References

Abdo, Z. and G. B. Golding. (2007). A step toward barcoding life: a model-based, decision-theoretic method to assign genes to preexisting species groups. *Systematic biology*, 56:44-56.

Altschul, S. F., W. Gish, W. Miller, E. W. Myers, and D. J. Lipman. (1990). Basic local alignment search tool. *Journal of molecular biology*, 215:403-410.

Barber, P. and S. L. Boyce. (2006). Estimating diversity of Indo-Pacific coral reef stomatopods through DNA barcoding of stomatopod larvae. *Proceedings of the Royal Society of London. Series B, Biological sciences*, 273:2053-2061.

Benson, D. A., I. Karsch-Mizrachi, D. J. Lipman, J. Ostell, and E. W. Sayers. (2011). GenBank. *Nucleic acids research*, 39:D32-37.

Casiraghi, M., M. Labra, E. Ferri, A. Galimberti, and F. De Mattia. (2010). DNA barcoding: a six-question tour to improve users' awareness about the method. *Briefings in bioinformatics*, 11:440-453.

CBOL Plant Working Group. (2009). A DNA barcode for land plants. *Proceedings of the National Academy of Sciences of the United States of America*, 106:12794-12797.

de Groot, G. A., H. J. During, J. W. Maas, H. Schneider, J. C. Vogel, and R. H. Erkens. (2011). Use of rbcL and trnL-F as a two-locus DNA barcode for identification of NW-European ferns: an ecological perspective. *PLoS one*, 6:e16371.

deWaard, J. R., A. Mitchell, M. A. Keena, D. Gopurenko, L. M. Boykin, K. F. Armstrong, M. G. Pogue, J. Lima, R. Floyd, R. H. Hanner, and L. M. Humble. (2010). Towards a global barcode library for Lymantria (Lepidoptera: Lymantriinae) tussock moths of biosecurity concern. *PLoS one*, 5:e14280.

Ebach, M. C. and C. Holdrege. (2005). DNA barcoding is no substitute for taxonomy. *Nature*, 434:697.

Eddy, S. R. 1998. Profile hidden Markov models. *Bioinformatics*, 14:755-763.

Evans, K. M., A. H. Wortley, and D. G. Mann. (2007). An assessment of potential diatom "barcode" genes (cox1, rbcL, 18S and ITS rDNA) and their effectiveness in determining relationships in Sellaphora (Bacillariophyta). *Protist*, 158:349-364.

Ewing, B. and P. Green. (1998). Base-calling of automated sequencer traces using phred. II. Error probabilities. *Genome research*, 8:186-194.

Fazekas, A. J., K. S. Burgess, P. R. Kesanakurti, S. W. Graham, S. G. Newmaster, B. C. Husband, D. M. Percy, M. Hajibabaei, and S. C. Barrett. (2008). Multiple multilocus DNA barcodes from the plastid genome discriminate plant species equally well. *PLoS one*, 3:e2802.

Frezal, L. and R. Leblois. (2008). Four years of DNA barcoding: current advances and prospects. *Infection, genetics and evolution,* 8:727-736.
Greenstone, M. H., D. L. Rowley, U. Heimbach, J. G. Lundgren, R. S. Pfannenstiel, and S. A. Rehner. (2005). Barcoding generalist predators by polymerase chain reaction: carabids and spiders. *Molecular ecology,* 14:3247-3266.
Gregory, T. R. (2005). DNA barcoding does not compete with taxonomy. *Nature,* 434:1067.
Hajibabaei, M., D. H. Janzen, J. M. Burns, W. Hallwachs, and P. D. Hebert. (2006). DNA barcodes distinguish species of tropical Lepidoptera. *Proceedings of the National Academy of Sciences of the United States of America,* 103:968-971.
Hebert, P. D., A. Cywinska, S. L. Ball, and J. R. deWaard. (2003a). Biological identifications through DNA barcodes. *Proceedings of the Royal Society of London. Series B, Biological sciences,* 270:313-321.
Hebert, P. D., E. H. Penton, J. M. Burns, D. H. Janzen, and W. Hallwachs. (2004a). Ten species in one: DNA barcoding reveals cryptic species in the neotropical skipper butterfly Astraptes fulgerator. *Proceedings of the National Academy of Sciences of the United States of America,* 101:14812-14817.
Hebert, P. D., S. Ratnasingham, and J. R. deWaard. (2003b). Barcoding animal life: cytochrome c oxidase subunit 1 divergences among closely related species. *Proceedings of the Royal Society of London. Series B, Biological Sciences,* 270 Suppl 1:S96-99.
Hebert, P. D., M. Y. Stoeckle, T. S. Zemlak, and C. M. Francis. (2004b). Identification of Birds through DNA Barcodes. *PLoS biology,* 2:e312.
Jaklitsch, W. M., M. Komon, C. P. Kubicek, and I. S. Druzhinina. (2006). Hypocrea crystalligena sp. nov., a common European species with a white-spored Trichoderma anamorph. *Mycologia,* 98:499-513.
Kaminuma, E., T. Kosuge, Y. Kodama, H. Aono, J. Mashima, T. Gojobori, H. Sugawara, O. Ogasawara, T. Takagi, K. Okubo, and Y. Nakamura. (2011). DDBJ progress report. *Nucleic acids research,* 39:D22-27.
Kerr, K. C., M. Y. Stoeckle, C. J. Dove, L. A. Weigt, C. M. Francis, and P. D. Hebert. (2007). Comprehensive DNA barcode coverage of North American birds. *Molecular ecology notes,* 7:535-543.
Kim, M. I., X. Wan, M. J. Kim, H. C. Jeong, N. H. Ahn, K. G. Kim, Y. S. Han, and I. Kim. (2010). Phylogenetic relationships of true butterflies (Lepidoptera: Papilionoidea) inferred from COI, 16S rRNA and EF-1alpha sequences. *Molecules and cells,* 30:409-425.
Kress, W. J. and D. L. Erickson. (2007). A two-locus global DNA barcode for land plants: the coding rbcL gene complements the non-coding trnH-psbA spacer region. *PLoS one,* 2:e508.
Kumar, N. P., A. R. Rajavel, R. Natarajan, and P. Jambulingam. (2007). DNA barcodes can distinguish species of Indian mosquitoes (Diptera: Culicidae). *Journal of medical entomology,* 44:1-7.
Lim, J., S. Y. Kim, S. Kim, H. S. Eo, C. B. Kim, W. K. Paek, W. Kim, and J. Bhak. (2009). BioBarcode: a general DNA barcoding database and server platform for Asian biodiversity resources. *BMC genomics,* 10 Suppl 3:S8.
Liu, J., M. Moller, L. M. Gao, D. Q. Zhang, and D. Z. Li. (2011). DNA barcoding for the discrimination of Eurasian yews (Taxus L., Taxaceae) and the discovery of cryptic species. *Molecular ecology resources,* 11:89-100.

Maiden, M. C., J. A. Bygraves, E. Feil, G. Morelli, J. E. Russell, R. Urwin, Q. Zhang, J. Zhou, K. Zurth, D. A. Caugant, I. M. Feavers, M. Achtman, and B. G. Spratt. (1998). Multilocus sequence typing: a portable approach to the identification of clones within populations of pathogenic microorganisms. *Proceedings of the National Academy of Sciences of the United States of America*, 95:3140-3145.

Meier, R., K. Shiyang, G. Vaidya, and P. K. Ng. (2006). DNA barcoding and taxonomy in Diptera: a tale of high intraspecific variability and low identification success. *Systematic biology*, 55:715-728.

Meyer, C. P. and G. Paulay. (2005). DNA barcoding: error rates based on comprehensive sampling. *PLoS biology*, 3:e422.

Mower, J. P., P. Touzet, J. S. Gummow, L. F. Delph, and J. D. Palmer. (2007). Extensive variation in synonymous substitution rates in mitochondrial genes of seed plants. *BMC evolutionary biology*, 7:135.

Mueller, R. L. (2006). Evolutionary rates, divergence dates, and the performance of mitochondrial genes in Bayesian phylogenetic analysis. *Systematic biology*, 55:289-300.

Nei, M. and S. Kuman. (2000). *Molecular Evolution and Phylogenetics*. Oxford University Press, New York.

Nielsen, R. and M. Matz. 2006. Statistical approaches for DNA barcoding. *Systematic biology*, 55:162-169.

Nilsson, R. H., M. Ryberg, E. Kristiansson, K. Abarenkov, K. H. Larsson, and U. Koljalg. 2006. Taxonomic reliability of DNA sequences in public sequence databases: a fungal perspective. *PLoS one*, 1:e59.

Pfenninger, M., C. Nowak, C. Kley, D. Steinke, and B. Streit. (2007). Utility of DNA taxonomy and barcoding for the inference of larval community structure in morphologically cryptic Chironomus (Diptera) species. *Molecular ecology*, 16:1957-1968.

Piredda, R., M. C. Simeone, M. Attimonelli, R. Bellarosa, and B. Schirone. (2011). Prospects of barcoding the Italian wild dendroflora: oaks reveal severe limitations to tracking species identity. *Molecular ecology resources*, 11:72-83.

Ran, J. H., P. P. Wang, H. J. Zhao, and X. Q. Wang. (2010). A test of seven candidate barcode regions from the plastome in Picea (Pinaceae). *Journal of integrative plant biology*, 52:1109-1126.

Ratnasingham, S. and P. D. Hebert. (2007). BOLD: The Barcode of Life Data System (http://www.barcodinglife.org). *Molecular ecology notes*, 7:355-364.

Rusch, D. B., A. L. Halpern, G. Sutton, K. B. Heidelberg, S. Williamson, S. Yooseph, D. Wu, J. A. Eisen, J. M. Hoffman, K. Remington, K. Beeson, B. Tran, H. Smith, H. Baden-Tillson, C. Stewart, J. Thorpe, J. Freeman, C. Andrews-Pfannkoch, J. E. Venter, K. Li, S. Kravitz, J. F. Heidelberg, T. Utterback, Y. H. Rogers, L. I. Falcon, V. Souza, G. Bonilla-Rosso, L. E. Eguiarte, D. M. Karl, S. Sathyendranath, T. Platt, E. Bermingham, V. Gallardo, G. Tamayo-Castillo, M. R. Ferrari, R. L. Strausberg, K. Nealson, R. Friedman, M. Frazier, and J. C. Venter. (2007). The Sorcerer II Global Ocean Sampling expedition: northwest Atlantic through eastern tropical Pacific. *PLoS biology*, 5:e77.

Schindel, D. E. and S. E. Miller. (2005). DNA barcoding a useful tool for taxonomists. *Nature*, 435:17.

Seifert, K. A., R. A. Samson, J. R. Dewaard, J. Houbraken, C. A. Levesque, J. M. Moncalvo, G. Louis-Seize, and P. D. Hebert. (2007). Prospects for fungus identification using CO1

DNA barcodes, with Penicillium as a test case. *Proceedings of the National Academy of Sciences of the United States of America*, 104:3901-3906.

Smith, M. A., B. L. Fisher, and P. D. Hebert. (2005). DNA barcoding for effective biodiversity assessment of a hyperdiverse arthropod group: the ants of Madagascar. *Philosophical transactions of the Royal Society of London. Series B, Biological sciences*, 360:1825-1834.

Stahls, G. and E. Savolainen. (2008). MtDNA COI barcodes reveal cryptic diversity in the Baetis vernus group (Ephemeroptera, Baetidae). Molecular phylogenetics and evolution 46:82-87.

Teletchea, F. (2010). After 7 years and 1000 citations: comparative assessment of the DNA barcoding and the DNA taxonomy proposals for taxonomists and non-taxonomists. Mitochondrial DNA 21:206-226.

Vences, M., M. Thomas, R. M. Bonett, and D. R. Vieites. (2005). Deciphering amphibian diversity through DNA barcoding: chances and challenges. *Philosophical transactions of the Royal Society of London. Series B, Biological sciences*, 360:1859-1868.

Venter, J. C., K. Remington, J. F. Heidelberg, A. L. Halpern, D. Rusch, J. A. Eisen, D. Wu, I. Paulsen, K. E. Nelson, W. Nelson, D. E. Fouts, S. Levy, A. H. Knap, M. W. Lomas, K. Nealson, O. White, J. Peterson, J. Hoffman, R. Parsons, H. Baden-Tillson, C. Pfannkoch, Y. H. Rogers, and H. O. Smith. (2004). Environmental genome shotgun sequencing of the Sargasso Sea. *Science*, 304:66-74.

Ward, R. D., T. S. Zemlak, B. H. Innes, P. R. Last, and P. D. Hebert. (2005). DNA barcoding Australia's fish species. *Philosophical transactions of the Royal Society of London. Series B, Biological sciences*, 360:1847-1857.

Wirth, T., D. Falush, R. Lan, F. Colles, P. Mensa, L. H. Wieler, H. Karch, P. R. Reeves, M. C. Maiden, H. Ochman, and M. Achtman. (2006). Sex and virulence in Escherichia coli: an evolutionary perspective. *Molecular microbiology*, 60:1136-1151.

Woese, C. R. (1996). Whither microbiology? Phylogenetic trees. *Current biology*, 6:1060-1063.

Yesson, C., R. T. Barcenas, H. M. Hernandez, M. De La Luz Ruiz-Maqueda, A. Prado, V. M. Rodriguez, and J. A. Hawkins. (2011). DNA barcodes for Mexican Cactaceae, plants under pressure from wild collecting. *Molecular ecology resources*, in publishing.

Zhou, J., M. E. Davey, J. B. Figueras, E. Rivkina, D. Gilichinsky, and J. M. Tiedje. (1997). Phylogenetic diversity of a bacterial community determined from Siberian tundra soil DNA. *Microbiology*, 143 (Pt 12):3913-3919.

Zhou, X., S. J. Adamowicz, L. M. Jacobus, R. E. Dewalt, and P. D. Hebert. (2009). Towards a comprehensive barcode library for arctic life - Ephemeroptera, Plecoptera, and Trichoptera of Churchill, Manitoba, Canada. *Frontiers in zoology*, 6:30.

10

Biological Data Modelling and Scripting in R

Srinivasan Ramachandran et al.[*]
*G.N. Ramachandran Knowledge Centre for Genome Informatics,
Institute of Genomics and Integrative Biology, Delhi,
India*

1. Introduction

In this age of Systems and Integrative Biology, development of high throughput genome sequencing techniques and other large-scale experimental methods, are generating large amount of biological data. Bioinformatics enables us to generate added value to these datasets in the form of annotation, classification and pattern extraction. These developments demand adequate storage and organization for further analysis.
In order to unravel the trends and patterns present in such diverse data sets, computational platforms with capability for carrying out integrative analysis are required for rapid analysis. R language platform is an example of one such platform allowing integrated rapid analysis process. The R is a High-level interpreted language suitable for developing new computational methods (R Development Core Team. 2010). Computational Biologists use R extensively because of the availability of numerous functions and packages including the well-known Bioconductor package (Gentleman et al., 2004). The rich inbuilt functions and the facility to write functions as well as object oriented programming facilities enable development of new packages for rapid analysis.

2. R platform

R is a programming language integrated with an R environment, facilitating easy and rapid data analysis with the help of its integrated suite of software facilities. Several computational biology packages have been developed in R language. Developing computational packages in R provides advantage as to carry out the analysis locally and also build further tools and scripts. Thus both new applications and extension of existing applications can be achieved. R helps accomplishment of complex tasks using simple scripts with the help of inbuilt suit of operators aiding in calculations. Also R environment provides graphical facilities for data analysis and display. Another major advantage of preparing datasets and computational biology tools in R is that a large set of statistical and mathematical tools can be applied on the datasets for analysis. R being an open source controlled by GNU General Public License allows future developments and customizations

---

[*]Rupanjali Chaudhuri, Srikant Prasad Verma, Ab Rauf Shah, Chaitali Paul, Shreya Chakraborty, Bhanwar Lal Puniya and Rahul Shubhra Mandal
G.N. Ramachandran Knowledge Centre for Genome Informatics, Institute of Genomics and Integrative Biology, Delhi, India

more widely. R is maintained by a core group of experts, thus ensuring its availability for long life. R in its repository also has a number of packages useful in various fields of biology. These packages help solve biological problems in well-structured manner saving time and money.

3. Data modeling for R

Data modeling for R involves identification of the datasets required for the corresponding problem undertaken. The data in the datasets needs to be structured into relevant rows and columns. For each field or column only one data type is allowed either character or numeric data type. Thereafter standardization or pre-processing of the data in datasets needs to be done. This involves checking the data for any inconsistencies- e.g., removal of blank cells by replacing with "Not known" or "None", checking header names for unwanted symbols like ?@$%*^ #/, checking columns for single data-type etc. The datasets may be then made into R object. Thus data modeling for R plays an important role to make data easily and properly read and operated with scripts in R platform. The data type in each column must conform to same format for all cells in that column.

4. S4 object oriented programming

S4 is the 4th version of S. The major development of S4 over S3 is the integration of functions, which allows considering S as an object oriented language. The object system in S4 provides a rich way of defining classes, handling inheritance, setting generic methods, validity checking and multiple dispatches. This allows development of easy to operate packages for rapid data handling and organized structured framework.

4.1 Setting class and reading data into S4 objects
Classes with specific representations are created in S4. Thereafter new object belonging to the set class may be created. Generic functions may also be made using object of the class:
1. setClass() is used to set the class of a data
2. new()is used to create objects of the class set
3. setGeneric() helps define generics
4. setMethods() is used to set methods

5. Decision tree

A decision tree (Maimon et al., 2005) is a tree like graph that a decision maker can create to help select the best amongst several alternative courses of action. Biological problems can be solved with help of well-structured and optimized algorithms. These algorithms can be represented in the form of decision trees to get better and clear understanding of the algorithm process followed to solve the biological problem.

6. Bioinformatics tools to retrieve biological data

Bioinformatics in its repository has a large number of tools developed to address diverse biological questions. These include investigating relationship between protein structure and function, immune response, development of potential vaccine candidates, modeling pathways, discovery of drug targets and drugs.

6.1 Immunoinformatics data

The immunoinformatics branch of bioinformatics deals with applying bioinformatics principles and tools to the molecular activities of the immune system. Immunoinformatics provides databases and predictive tools, useful to fetch data on cells of immune system. This data is termed immunological data and can be broadly split into epitope data and allergen data. This data is useful for aiding in vaccine discovery, referred to as computer aided vaccine design. An important aim here is antigen identification or identification of epitopes capable of eliciting immune response. There are various immunoinformatics databases available for aiding this process (Chaudhuri et al., 2008; Chaudhuri et al., 2011; Vivona et al., 2008).

An epitope, also known as 'antigenic determinant' is a surface localized part of antigen capable of eliciting an immune response. A B-cell epitope is region of the antigen recognized by soluble or membrane bound antibodies. B-cell epitopes are further classified as either linear or discontinuous epitopes. Linear epitope is a single continuous stretch of amino acids within a protein sequence, whereas epitopes whose residues are distantly placed in the sequence but are brought together by physico-chemical folding are termed as discontinuous epitopes.

T cell epitope is a short region presented on the surface of an antigen-presenting cell, where they are bound to MHC molecules. These epitopes can be characterized into two types based on their recognition by either MHC Class I molecule or Class II molecule.

Epitope prediction tools form the backbone of immunoinformatics. The main aim of these tools is to aid in reliable epitope identification. Various sophisticated T cell epitope prediction tools have been developed which help successful epitope prediction. Some of these algorithms are based on artificial neural networks and weight matrices such as NetMHC (Lundegaard et al., 2008), predictive IC(50) values IEDB-ARB method (Bui et al., 2005; Zhang et al., 2008), predicted half-time of dissociation Bimas (Parker et al., 1994), quantitative matrices ProPred (Singh et al., 2001). Reliable and accurate B-cell epitope prediction is still in development although we have some tools such as ABCpred (Saha et al., 2006) and BcePred (Saha et al., 2007). These tools help build the epitope data from protein sequences.

Allergen identification holds major importance in vaccine discovery problem, as it is desirable that a candidate vaccine is non-allergic. Allergens are substances (proteins, carbohydrates, particles, pollengrains etc.) to which the body mounts a hypersensitive immune response typically of Type I.

Various tools of immunoinformatics have been developed with aim to predict allergenic proteins. AlgPred (Saha et al., 2006) allows prediction of allergens through either singly or in combination of support vector machine, motif-based method, and searching the database of known IgE epitopes. Allermatch (Fiers et al., 2004) performs BLAST search against allergen representative peptides using a sliding window approach. The data fetched constitute allergen data. The building of Dataclasses with their representations is described in Figures 1-4.

6.1.1 Identification of potential immunogens useful as vaccine candidates

Immunogen is a substance capable of eliciting an immune response. It possesses epitopes, which binds to the B cells or T cells to elicit the response. To identify protein immunogens useful as vaccine candidates, bioinformatics approach may be undertaken. There are various B-cell and T-cell epitope prediction tools available as mentioned in the previous section. These algorithms provide prediction of the epitopes present in the submitted protein sequence. Each prediction comes with associated score representing the confidence of prediction. A cutoff score can be set to select the high scoring epitopes and subsequently proteins can be identified with high scoring epitopes. Thus the filtered orfids of proteins with high scoring B-cell and T-cell epitopes can be selected. The individual results of orfids may be analyzed by using the 'intersect' operator of R to get the final list of orfids representing the proteins meeting conditions of multiple features. It is desirable for the candidate vaccine to be non-allergic. Allergen data for the proteins may be fetched using allergen prediction immunoinformatics tools to obtain list of non-allergic proteins. Thus list of non-allergen proteins with high scoring B-cell and T-cell epitopes may be obtained. B-cell and T-cell data have been captured as **Secondlayer** data. As an example from the **Firstlayer** data certain sub problems to target a potential adhesin vaccine candidate can be stated as- the protein should be an adhesin, the protein should not be intracellularly located, it should not have similarity to human reference proteins, it should not have more than one transmembrane helix thereby facilitating proper cloning and expression. The set of proteins fulfilling all the **Firstlayer** conditions can be intersected with the set of non-allergen proteins. This whole process is depicted as decision tree (Figure 5). Similarly the decision tree describing the steps for obtaining proteins with high scoring B-cell and T-cell epitopes is shown in Figure 6.

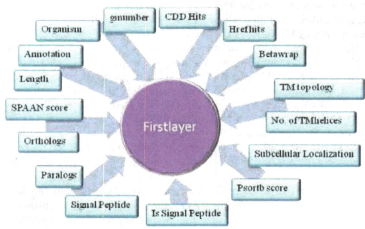

```
setClass("FirstLayer", representation(ginumber = "numeric", annot = "character", length = "numeric", spaanscore = "numeric", paralogs = "character", omcl = "character", signalp = "numeric", is_signalp = "character", psortbscore = "numeric", subclllocal = "character", tmhelices = "numeric", topotmhelix = "character", betawrap = "character", Hrefhits = "character", cddhits = "character"))
```

```
readdata.firstlayer<-
function(xz){xa<-
readLines(con = xz);
tempy<- NULL;for (i in
seq ( along =
xa)){tempx<-
unlist(strsplit(xa[i],"\t"))
;tempy<- c(tempy,
new("FirstLayer",
ginumber =
as.numeric(tempx[1]),
annot = tempx[2], length
= as.numeric(tempx[3]),
spaanscore =
as.numeric(tempx[4]),
paralogs = tempx[5],
omcl = tempx[6],
signalp =
as.numeric(tempx[7]),
is_signalp = tempx[8],
psortbscore =
as.numeric(tempx[9]),
subcelllocal =
tempx[10], tmhelices =
as.numeric(tempx[11]),
topotmhelix =
tempx[12], betawrap =
tempx[13], Hrefhits =
tempx[14], cddhits =
tempx[15] )) };
return(tempy)}
```

Fig. 1. Representation of S4 Class "FirstLayer" and the R scripts to accomplish the construction.

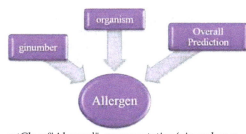

setClass("Algpred", representation(ginumber = "numeric" , organism="character", ovpr= "character",IGEPred="character",IgEepitope="character",Seqmatched="character",position="numeric",PID="numeric", MASTRESULT="character", SVMPRED="character",SVMScore="numeric",SVMThold="numeric"
,SVMPPV="character",SVMNPV="character",SVMDipepPRED="character", SVMDipepScore="numeric",SVMDipepThold="numeric",SVMDipepPPV="character",SVMDipepNPV="character",BLASTPred="character",HitARPs="character")); setClass("Allermatch", representation(ginumber = "numeric" , organism="character", prediction ="character",hit_no= "numeric", db ="character", allermatch_id ="character",best_nit_index ="numeric",no_hits_ident_gt35 ="numeric",perc_hits_gt35 ="numeric", perc_ident ="numeric", seq_len_fasta_aligned= "numeric",external_link ="character", link_db ="character",genus_name ="character", spc_name="character"))

readdata.algpred<- function(xz){xa<- readLines(con = xz); tempy<- NULL;for (i in seq (along = xa)){tempx<- unlist(strsplit(xa[i],"\t"));tempy<- c(tempy, new("Algpred", ginumber = as.numeric(tempx[1]), organism = tempx[2], ovpr= tempx[3] ,IGEPred=tempx[4] ,IgEepitope=tempx[5] ,Seqmatched= tempx[6],position= as.numeric(tempx[7]) ,PID= as.numeric(tempx[8]), MASTRESULT=tempx[9], SVMPRED=tempx[10], SVMScore= as.numeric(tempx[11]),SVMThold= as.numeric(tempx[12]) ,SVMPPV= tempx[13],SVMNPV= tempx[14],SVMDipepPRED= tempx[15], SVMDipepScore= as.numeric(tempx[16]),SVMDipepThold= as.numeric(tempx[17]),SVMDipepPPV= tempx[18],SVMDipepNPV= tempx[19],BLASTPred= tempx[20],HitARPs= tempx[21]))};return(tempy)}
readdata.allermatch<- function(xz){xa<- readLines(con = xz); tempy<- NULL;for (i in seq (along = xa)){tempx<- unlist(strsplit(xa[i],"\t"));tempy<- c(tempy, new("Allermatch",ginumber = as.numeric(tempx[1]), organism = tempx[2], prediction= tempx[3], hit_no = as.numeric(tempx[4]),db= tempx[5], allermatch_id= tempx[6], best_nit_index= as.numeric(tempx[7]), no_hits_ident_gt35= as.numeric(tempx[8]),perc_hits_gt35= as.numeric(tempx[9]),perc_ident= as.numeric(tempx[10]), seq_len_fasta_aligned= as.numeric(tempx[11]), external_link= tempx[12], link_db= tempx[13], genus_name= tempx[14], spc_name= tempx[15]))};return(tempy)}

Fig. 2. General representation of S4 Class for Allergen data. The script for reading the data in is given for AlgPred. Similarly the data can be read for Allermatch class with appropriate data representation.

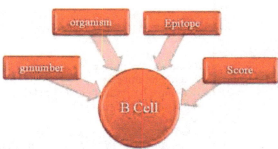

setClass("Bcepred", representation(ginumber = "numeric",organism= "character", property= "character", sequence= "character", length= "numeric"));setClass("ABCpred", representation(ginumber = "numeric" , organism="character", rank="numeric" , sequence="character", position="numeric", score="numeric"))

readdata.bcepred<- function(xz){xa<- readLines(con = xz); tempy<- NULL;for (i in seq (along = xa)){tempx<- unlist(strsplit(xa[i],"\t"));tempy<- c(tempy, new("Bcepred", ginumber = as.numeric(tempx[1]), organism = tempx[2], property = tempx[3], sequence = tempx[4], length = as.numeric(tempx[5])))}; return(tempy)}
readdata.abcpred<- function(xz){xa<- readLines(con = xz); tempy<- NULL;for (i in seq (along = xa)){tempx<- unlist(strsplit(xa[i],"\t"));tempy<- c(tempy, new("ABCpred", ginumber = as.numeric(tempx[1]), organism = tempx[2], rank = as.numeric(tempx[3]), sequence = tempx[4], position = as.numeric(tempx[5]), score= as.numeric(tempx[6])))};return(tempy)}

Fig. 3. General representation of S4 Class for B Cell epitope data along with R scripts.

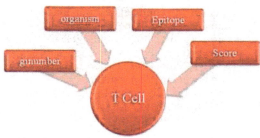

setClass("Propred", representation(ginumber = "numeric",organism= "character", Allele= "character", Rank= "numeric", Sequence= "character", Position= "numeric", Score= "numeric"))

readdata.propred <-function(xz){xa<- readLines(con = xz); tempy<- NULL;for (i in seq (along = xa)){tempx<- unlist(strsplit(xa[i],"\t"));tempy<- c(tempy, new("Propred", ginumber = as.numeric(tempx[1]), organism = tempx[2],Allele= tempx[3], Rank= as.numeric(tempx[4]), Sequence= tempx[5], Position= as.numeric(tempx[6]), Score= as.numeric(tempx[7])))};return(tempy)}

Fig. 4. General representation of S4 Class for T Cell epitope data.

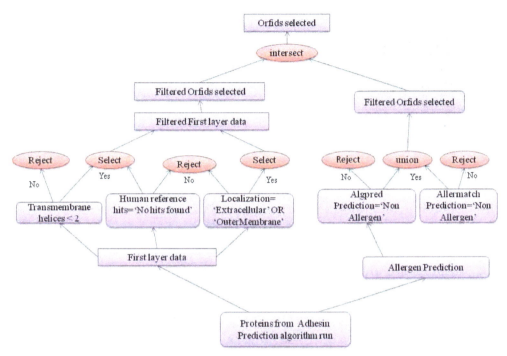

Fig. 5. Decision tree to identify non-allergen proteins fulfilling all first layer conditions. The R scripts are shown in the following two boxes.

```
S4 Methods
setGeneric("getfl_filtered",function(object)
standardGeneric("getfl_filtered"));setMethod("getfl_filtered","FirstLayer",function(obj
ect){if ((object@tmhelices < 2) && (object@Hrefhits== "No Hits found") &&
((object@subcelllocal == "Extracellular") || (object@subcelllocal ==
"OuterMembrane"))) {return (object@ginumber)}else{return(FALSE)}})
setGeneric("nonallergen_algpred",function(object)
standardGeneric("nonallergen_algpred"));setMethod("nonallergen_algpred","Algpre
d",function(object){if
( object@ovpr == "Non Allergen") {return (object@ginumber)}else{return(FALSE)}})
setGeneric("nonallergen_allermatch",function(object)
standardGeneric("nonallergen_allermatch"));setMethod("nonallergen_allermatch","Al
lermatch",function(object){if
( object@prediction == "Non Allergen") {return
(object@ginumber)}else{return(FALSE)}})
```

```
R Scripts
res1<- sapply(ecalgpred,nonallergen_algpred); res2<-
sapply(ecallermatch,nonallergen_allermatch)
resA<- union(res1,res2); resB <- sapply(ecflnew,getfl_filtered); resC<-
intersect(resA,resB)
```

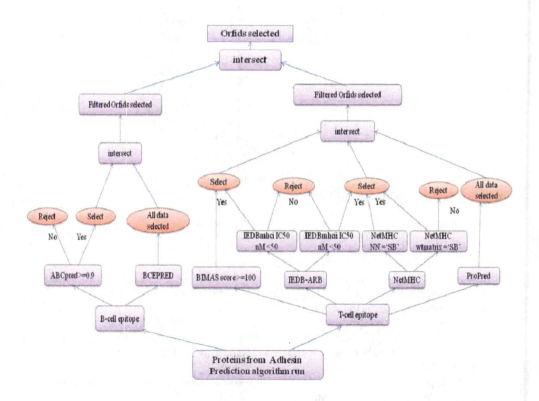

Fig. 6. Decision tree to identify high scoring B cell and T cell epitopes. The R scripts follow in the next two boxes.

S4 Methods
setGeneric("getgibce", function(object)
standardGeneric("getgibce"));setMethod("getgibce","Bcepred",function(object){
object@ginumber})
setGeneric("getgi_abcepitopes", function(object,x)
standardGeneric("get_abc_epi_gi"));setMethod("get_abc_epi_gi","ABCpred",function(object,x){if (object@score >= x) {return
(object@ginumber)}else{return(FALSE)}})
setGeneric("getgipropred", function(object)
standardGeneric("getgipropred"));setMethod("getgipropred","Propred",function(object){object@ginumber})
setGeneric("getgi_bimasepitopes",function(object,x)
standardGeneric("getgi_bimasepitopes"));setMethod("getgi_bimasepitopes","Bimas",function(object,x){if (object@Score >= x) {return
(object@ginumber)}else{return(FALSE)}})
setGeneric("getgi_NetMHCNNepitopes",function(object)
standardGeneric("getgi_NetMHCNNepitopes"));setMethod("getgi_NetMHCNNepitopes","NetMHCneuralnet",function(object){if (object@Bind_level == "SB")
{return (object@ginumber)}else{return(FALSE)}})
setGeneric("getgi_NetMHCwtepitopes",function(object)
standardGeneric("getgi_NetMHCwtepitopes
"));setMethod("getgi_NetMHCwtepitopes
","NetMHCwtmatrix",function(object){if (object@Bind_level == "SB") {return
(object@ginumber)}else{return(FALSE)}})
setGeneric("getgi_iedbmhciepitopes", function(object,x)
standardGeneric("getgi_iedbmhciepitopes
"));setMethod("getgi_iedbmhciepitopes ","IEDB_mhci",function(object,x){if (
object@IC50 < x) {return (object@ginumber)}else{return(FALSE)}})
setGeneric("get_iedb_mhciiepi_gi", function(object,x)
standardGeneric("get_iedb_mhciiepi_gi"));setMethod("get_iedb_mhciiepi_gi","IEDB_mhcii",function(object,x){if (object@IC50 < x) {return
(object@ginumber)}else{return(FALSE)}})

```
R Scripts
resabc<- sapply(eclabc,getgi_abcepitopes,0.9); nrresABC<- union(resabc,resabc);
resbce<- sapply(eclbce,getgibce); nrresBCE <- union(resbce,resbce); resfl1<-
intersect(nrresABC,nrresBCE); resbimas <-
sapply(ecbimas,getgi_bimasepitopes,100); nrresBIMAS <-
union(resbimas,resbimas); resiedb<-
sapply(ec_iedb_mhci,getgi_iedbmhciepitopes,50); resiedbmhc1 <-
sapply(ec_iedb_mhcii, getgi_iedbmhciiepitopes, 50); nrresIEDBMHC2 <-
union(resiedbmhc1,resiedbmhc1); resnetmhcnn<-
sapply(ecNetMHCneuralnet,getgi_NetMHCNNepitopes); nrresNETMHCNN <-
union(resnetmhcnn,resnetmhcnn); resnetmhcwtmat <-
sapply(ecNetMHCwtmatrix,getgi_NetMHCwtepitopes); nrresNETMHCWTMAT
<- union(resnetmhcwtmat,resnetmhcwtmat); respropred<-
sapply(ecpropred,getgipropred); nrresPROPRED<-
union(respropred,respropred); nr1<- intersect(nrresBIMAS,nrresIEDB); nr2<-
intersect(nr1,nrresIEDBMHC2); nr3<- intersect(nr2,nrresNETMHCNN); nr4<-
intersect(nr3,nrresNETMHCWTMAT); nr5<- intersect(nr4,nrresPROPRED);
selectedgis<- intersect(resfl1,nr5)
finalgis<- intersect(selectedgis,resC)
```

6.2 Systems biology data

Systems biology deals with a system-level understanding of biological systems. A system can be defined by a set of interacting entities, which are linked to each other by direct and indirect interactions. A biological system is a very complex network, which cannot be described by reductionist's approach because it gives us a limited knowledge of a particular gene or protein that is insufficient to understand the complex behavior of a biological network. There is a need to integrate all the knowledge and comprehend new networks, which provide the overall picture of a system. These inferred networks can be used for further computational analysis and if found promising, can be validated through experiments. System level understanding requires the integration of experimental and computational biology. Modeling is the best method to represent a pathway and is the easiest way to understand a complex network. A network is modeled as a graph, which is the formal mathematical representation of the network and consists of nodes and edges. The network can be shown diagrammatically by using *classical graph theory*. All type of pathways (e.g. Gene regulatory network, signal transduction and metabolic pathways) can be modeled using various modeling techniques. A modeler uses two types of approaches- Data driven pathway modeling and Knowledge based pathway modeling, depending on the presence/absence of sufficient literature (Viswanathan et al., 2008). If the knowledge is limited, data driven pathway modeling becomes the best choice. These modeling techniques are also known as qualitative (Data driven) and quantitative (Knowledge driven) modeling approaches. Data driven pathway modeling requires the DNA microarray data set. For example, the Gene Regulatory network (GRN) can be inferred by using logical networks like Boolean networks, probabilistic Boolean network and dynamic Bayesian networks (Li et al., 2007).

A quantitative model describes a system with a set of mathematical equations. Recently, many software tools have been developed for quantitative modeling of biological systems. We know that all physiochemical reactions follow a physical or chemical principle. For example a given enzyme catalysis reaction may follow the Michaelis Menten kinetics (Nelson et al, 2000). Thus, every reaction in kinetic model is represented in kinetic equation, which is then solved by the ordinary differential equation. In other words, a model is represented as a system of ODEs (Ordinary Differential Equations) for each of the reactions involved in the pathway (Tyson et al., 2001). If kinetic parameters are available, ODE based modeling becomes the best tool to understand dynamics of network.

There are variety of bioinformatics tools available for modeling systems in many platforms. (Table 1 and Table 2)

Task	Tools	Web address
Model construction	CellDesigner	http://www.celldesigner.org/
	Jarnac	http://sys-bio.org/
	Jdesigner	http://sys-bio.org/
	Gepasi	http://www.gepasi.org/
Simulation	CellDesigner	http://www.celldesigner.org/
	COPASI	http://www.copasi.org/
	Gepasi	http://www.gepasi.org/
	SBaddon (MatLab tool)	http://www.mathworks.com/
Model Analysis	MatLab,	http://www.mathworks.com/
	R- environment	http://www.r-project.org/

Table 1. Bioinformatics tools for systems modeling in different platforms.

Package Name	Application
BoolNet	Generation, reconstruction, simulation and analysis of synchronous, asynchronous, and probabilistic Boolean networks
odesolve	Solver for ordinary differential equations
lpSolve	Interface to solve linear/integer programs
nlme	Linear and non-linear mixed effect model
SBML-R	SBML are R interface analysis tool

Table 2. Tools for systems modeling in R platform

6.2.1 Examining the expression pattern of genes in clinical strains, an example

This process is initiated by first collecting the microarray data from public repository. Next data normalization needs to be done. Log (base=10) transformed data can be used to normalize by using classical Z-score transformation method (Cheadle et al., 2003). Z-score reflects the relative expression condition of the genes. On the basis of z-score values we can categorize genes in many categories like highly expressed, moderately expressed and genes with low expression. We can also filter those genes having the z-score values higher than given cutoff in all samples or strains. The consistency of expression across the different samples or strains can be explained using Heatmap. R scripts may be used to obtain genes having z-score above 1 which would provide genes which are highly expressed. Heatmap can be generated by using the R scripts. These heatmaps are false color image and very helpfull for visual comparison of different datasets. Dendrogram can be added on rows and columns by defining the heatmap arguments. The function Heatmap is provided by Bioconductor (Gentleman et al, 2004).

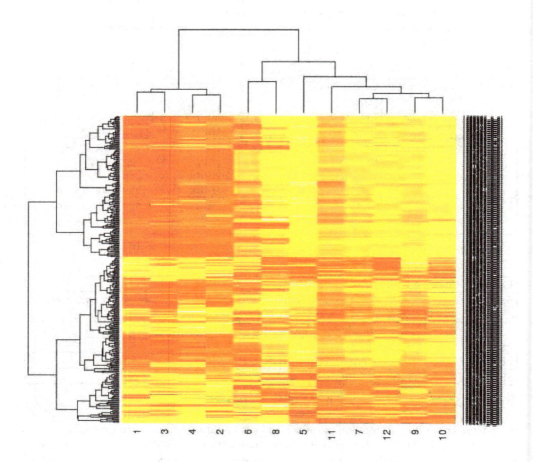

Fig. 7. Heat map of all probesets with z-score greater than 1.0 in all 12 samples. Red – Lower limit, Yellow - Upper limit gene expression Zscores. The sample ids are labelled below.

Biological Data Modelling and Scripting in R 213

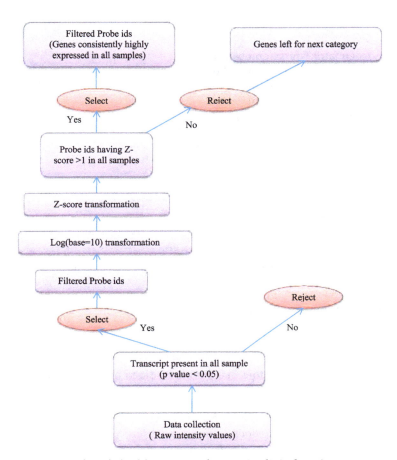

Fig. 8. Decision tree to identify highly expressed genes in clinical strains.

setClass("ZscoreEcoli",representation(probe_id = "character", expressionmat = "matrix"))

readdata.ZscoreEcoliExp<- function(xz){xa<-readLines(con = xz); tempy<- NULL;for (i in seq (along = xa)){tempx<- unlist(strsplit(xa[i],"\t"));tempy<- c(tempy, new("ZscoreEcoli", probe_id = tempx[1], expressionmat = matrix(c(as.numeric(tempx[2]),as.numeric(tempx[3]),as.numeric(tempx[4]),as.numeric(tempx[5]),as.numeric(tempx[6]),as.numeric(tempx[7]),as.numeric(tempx[8]),as.numeric(tempx[9]),as.numeric(tempx[10]),as.numeric(tempx[11]),as.numeric(tempx[12]),as.numeric(tempx[13])), nrow=1,ncol=12))) }; return(tempy)}
[11]),as.numeric(tempx[12]),as.numeric(tempx[13])), nrow=1,ncol=12))) }; return(tempy)}

Fig. 9. Representation of S4 Class "ZscoreEcoli".

S4 Methods
setGeneric("getmatrix", function(object,x))
standardGeneric("getmatrix"));setMethod("getmatrix","ZscoreEcoli",function(object,x
){tempmat<- object@expressionmat; tempnew<- NULL; if ((tempmat[1,1] > x) &&
(tempmat[1,2] > x) && (tempmat[1,3] > x) && (tempmat[1,4] > x) && (tempmat[1,5]
> x)&& (tempmat[1,6] > x)&& (tempmat[1,7] > x)&& (tempmat[1,8] > x)&&
(tempmat[1,9] > x)&& (tempmat[1,10] > x)&& (tempmat[1,11] > x)&&
(tempmat[1,12] > x)) {tempnew <- new("ZscoreEcoli", probe_id = object@probe_id,
expressionmat = tempmat);return(tempnew)} else return(0)})

R Scripts
eczscore<- readdata.ZscoreEcoliExp("zscoreEcoli")
EcoliMat <- sapply(eczscore,getmatrix,1); EcoliMatFinal <- setdiff(EcoliMat,0)
mymat <- NULL; for (j in seq(along = EcoliMatFinal)){tempomat<-
EcoliMatFinal[[j]]@expressionmat; rownames(tempomat) <-
EcoliMatFinal[[j]]@probe_id; mymat<- rbind(mymat,tempomat)}
heatmap(mymat)

6.2.2 Identifications of the attractors in a simple Boolean network using BoolNet package

Biological entities can have 2 possible logical states ON or OFF i.e. transcription of gene being either ON or OFF, protein is either Present or Absent etc. A system is more intuitively understandable by logical assumptions. Mainly Boolean logical network is used for the gene regulatory networks. Here we have implemented it on a more simple metabolic reactions network. Here we have chosen 3 reactions of genes involved in metabolic pathways of *Mycobacterium tuberculosis*, which are consistently highly expressed in 12 different strains (Gao et al., 2005). This Boolean network consists of 8 genes. Simple rules are written for reaction by using AND and OR Boolean operators. Attractors are the points in a network towards which the system is evolved. Attractors can be steady states or cycles. These are the states where system resides most of the time (Müssel et al., 2010).
###########
Boolean network with 8 genes
Involved genes:
nad coa oaa pyr sdhlam accoa succoa cit
Transition functions:
nad = nad
coa = coa
oaa = oaa
pyr = pyr
sdhlam = sdhlam
accoa = (coa & nad & pyr) | (cit & coa)

succoa = (coa & sdhlam)
cit = (accoa & oaa)
##########
Abbreviations
NAD = NAD, COA = Coenzyme-A, OAA = oxaloacetate, PYR = pyruvate, SDHLAM= S-adenosyl-L-methionine, ACCOA= acetyl-CoA, CIT = citrate, SUCCOA = succinyl-coA
Description of network: -
1. AcetylCoA is formed by 2 reactions: Coenzyme A and NAD and Pyruvate OR Citrate and Coenzyme A
2. SuccinylCoA is formed by Coenzyme A and S-adenosyl-L-methionine
3. Citrate is formed by Acetyl-CoA and Oxaloacetate
4. Species considered as constant (whose rules are not defined) e.g. NAD, OAA, PYR, SDHLAM and COA

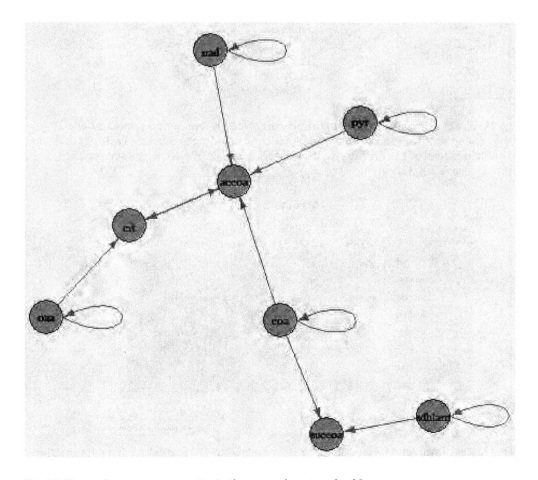

Fig. 10. Dependency among species in the example network of 8 genes

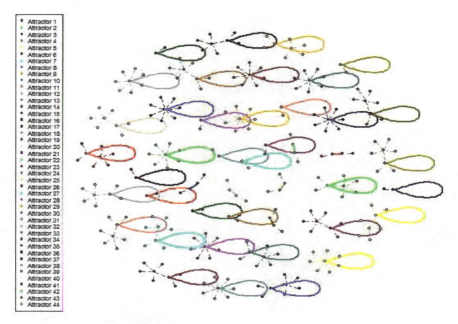

Fig. 11. Plot of different attractors in the network. Node is representing state and line representing state transition. Colour in plot is corresponding to different basin of attraction. We obtained total of 44 attractors for the network of which 39 have single state and 5 attractors have 2 states.

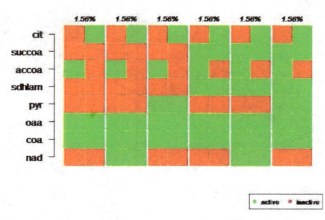

Fig. 12. Plot of Attractors with 2 states (from subsets 39 to 44). Red colour given for inactive genes and green colour is given for active genes. Each attractor given in plot is contributing 1.56% in the network.

Even though this is simple reaction network, yet it is giving the view about the states where system can reside most of the time i.e. attractors in the network. We can study the knockout and over expression in a complex network. Robustness study of the network is also possible by studying network behavior through knocking out genes.

6.3 Chemoinformatics data

The Chemoinformatics branch is the interface between Computer applications and Chemistry and deals with problems of the field of the chemistry. Chemoinformatics concentrates on molecular modelling, chemical structure coding and searching, data visualization etc.

Molecular modelling involves the use of theoretical methods and computational techniques to model or mimic the behavior of molecules. It helps reduce the complexity of the system, allowing many more particles (atoms) to be considered during simulations. Data visualization is the study of the visual representation of data by graphical means.

Chemoinformatics is useful specially to solve drug discovery related problems. Drug-like or Lead identifications are done through various in-silico methods in chemoinformatics. Drug-like compounds refer to the compounds, which follow the lipinski's rule and have structural similarities with the known drugs and bind to the active site of the target but have not been tested in laboratory. There are also various databases available helping in this direction.

Functions	Tools	Links
Databases for searching Known Inhibitors	Pubchem Pubmed Drug Bank	http://pubchem.ncbi.nlm.nih.gov/search/search.cgi http://www.ncbi.nlm.nih.gov/pubmed http://www.drugbank.ca/
Molecular Visualization	Pymol Rasmol SwissPDBviewer	http://www.pymol.org http://rasmol.org http://spdbv.vital-it.ch/
Structres drawn StructuresViewed	Marvin Sketch Chemmine	http://www.chemaxon.com/products/marvin/marvinsketch http://bioweb.ucr.edu/ChemMineV2
File Format Translator	Smile Translator OpenBabel	http://cactus.nci.nih.gov/translate/ http://openbabel.org/wiki/Main_Page
Drug Designing	AutoDock GOLD	http://autodock.scripps.edu/ http://www.ccdc.cam.ac.uk/products/life_sciences/gold/
Toxicity prediction	Toxtree	http://toxtree.sourceforge.net/
Molecular dynamics simulation	GROMACS	http://www.gromacs.org/

Table 3. Softwares used in Cheminformatics.

Package Name	Principle	Application
ChemmineR	Uses Tanimoto coefficient as the similarity measure	Atom-pair descriptors calculation with the help of functions included within it, 2D structural similarity searching, clustering of compound libraries, visualization of clustering results and chemical structures.
Rcdk	Allow an user to access functions of CDK (JAVA library for Chemiinformatics) on R platform	Reading molecular file formats, performing ring perception, aromaticity detection to fingerprint generation and determining molecular descriptors of various properties of Drug-Like compounds
Rpubcem	Uses various functions for data retrieval	Helps access the datas and assays of compounds from pubchem
ic50	Helps determine the efficiency of a newly found drug like molecule	Calculates IC50
Bio3D	Analysis of protein structure and sequence data	Protein structure analysis, comparative analysis with different proteins, aligns protein sequences.

Table 4. Tools useful in Chemoinformatics in R

6.3.1 Identification of drug-like compounds

In this era, with the rise in number of infectious life threatening diseases due to clever change at sequence level of the pathogenic organism, discovery of new potential drugs holds immense importance. In this direction various in-silico chemoinformatics tools mentioned in Table 3 are helpful. In the problem regarding identification of Drug-like compounds the initial step would include literature search for known inhibitors of the disease target. Similarly a database of publicly available drug like molecules is obtained from various databases like ZINC (Irwin et al., 2005), NCI (Voigt et al., 2001) database etc. Thereafter Marvin sketch (MarvinSketch 5.3.8, 2010) may be used to draw the known inhibitors and their analogs, which need to be saved in file formats: Structure Data Format (SDF) and Simplified Molecular Input Line Entry Specification (SMILES). Using ChemmineR package (Cao et al, 2008) on R platform similarity search of Known inhibitor with the database of publicly available drug-like molecules is done using Tanimoto Coefficient where a desirable cutoff score e.g. 0.6 may be used as cutoff score to obtain a list of similar lead compounds. This list corresponds to a number of compounds similar to the known inhibitors. The properties of these similar compounds are calculated with another package of R called rcdk (Guha et al, 2007). The Lipinski's rule of five can be applied to further shortlist. The solubility of these compounds can be analysed by pHSol 1.0 Server (Hansen et al, 2006). For analyzing protein-ligand binding, docking is done using AutoDock software (Goodsell et al, 1996). If the ligand binds to the active site, only then it may have potential to interfere with protein function thereby eligible for further testing. Thereafter to test its potential further, energy minimization and simulation are done with various softwares eg. GROMACS (Hess et al., 2008). The ligand fulfilling all the desirable drug like quality and showing good stability over a reasonable time period of simulations (5-10 nanoseconds) may be selected as a candidate for testing. This process is represented as decision tree (Figure 13).

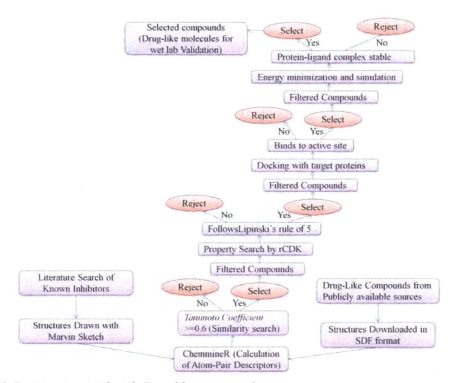

Fig. 13. Decision tree to identify Drug like compounds.

```
readdata.MolDescriptors=
function(filename){x=readLines(
con=filename);temp=NULL;obj=
NULL;for(i in
2:length(x)){temp=unlist(strsplit(
x[i],"\t"));print
(temp);obj=c(obj,new("MolDescri
ptors",Mol_id=as.numeric(temp[
1]),TPSA=as.numeric(temp[2]),n
HBAcc=as.numeric(temp[3]),nH
BDon=as.numeric(temp[4]),nRot
B=as.numeric(temp[5]),LipinskiF
ailures=as.numeric(temp[6]),MW
=as.numeric(temp[7]),XLogP=as.
numeric(temp[8]),SMILE=temp[
9]))};return(obj);}
```

setClass(Class="MolDescriptors",representation=representation
(Mol_id="numeric",TPSA="numeric",nHBAcc="numeric",nHB
Don="numeric",nRotB="numeric",LipinskiFailures="numeric",
MW="numeric",XLogP="numeric",SMILE="character"))

Fig. 14. Representation of S4 Class "MolDescriptors" with R scripts to accomplish the construction.

> S4 method to get filtered molecular Descriptor Data
> setGeneric('FilterMols",function(obj,x)standardGeneric("FilterMols"));
> setMethod("FilterMols",
> "MolDescriptors",function(obj,x){if(obj@LipinskiFailures==x){return(obj);}else{return(0);}})

> Example R Scripts to get filtered molecular Descriptor Data
> result=sapply(mol,FilterMols,1); result_final=setdiff(result,"0")

6.4 Text mining

During last few decades there has been enormous increase in the size of research data in the form of scientific articles, abstracts and books, online databases and many more. This text data may be structured or unstructured and need of hour is to mine for useful information from text data.

Thus text mining has evolved widely as an interdisciplinary discipline using methods from computer sciences, linguistics and statistics. R provides intelligent ways in accessing and integrating the treasure of information hidden in the scientific journals, papers and other electronic media. It is most often used to perform statistical and data mining analyses and is best known for its ability to analyze structured data. Majority of people read only abstracts of papers so as to save time and avoid in going into details of irrelevant articles. The **tm** R-package provides complete platform that efficiently processes various text documents to extract useful information (Feinerer et al, 2008). The database backend support also minimizes the memory demands to handle very large data sets in R. It accepts text data either from local database or directly from online database.

6.4.1 An example of text mining PubMed abstracts to get frequently appearing gene symbols and drug names

We downloaded abstracts for PubMed query "Glioma" from PubMed. We formatted these abstracts in R so as to get a file containing PMIDs, Titles, Abstract Text and Journal Name in separate columns. We create the S4 class object of this Abstract file. Now if we have to search for any abstract containing a pattern or word of our interest we use *regexpr* function in R to get all those abstracts e.g. pathway.

From these filtered Glioma pathway abstracts a Corpus (collection of large text) is generated using tm package function. This Corpus is subjected to Stemming, Stop word removal, common English word removal using R libraries rJava, RWeka (Hornik et al, 2009), RWekajars, slam, Snowball (Hornik, 2009), Corpora. We can set other controls as well. Now we can create term document matrix from this Corpus, containing all terms as rows and abstracts as columns. From this term document matrix we can extract the terms within different frequency ranges i.e., their number of occurrences in each abstract indicating their importance. Thus we can predict that terms which have higher frequency of occurrence in the matrix are more important and are related in some sense. We have filtered all the terms with frequency of their occurrence greater than 5.

These terms were 'intersect' with HUGO Gene Nomenclature Committee data set (HGNC) (Bruford et al, 2008) and drug names from DrugBank (Knox, et al 2011). We thus get the list of all the gene symbols, gene names, gene aliases and drug names in the said frequency range from term document matrix. We now determine the total count of their occurrence in matrix so as to rank these extracted genes and drugs. We can make clusters from these data and can find gene-drug, gene-gene, and drug-gene interactions in different ranges of correlation. The decision tree is shown in Figure 15. The dataclasses with their representations is described in Figure 16.

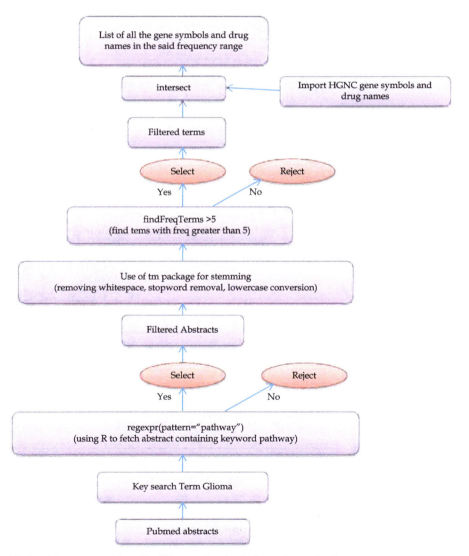

Fig. 15. Decision tree to identify **Glioma** that include pathway in the text.

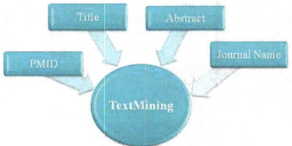

readdata.textmining<-function(xz){xa<- readLines(con = xz);tempy<- NULL;for (i in seq (along = xa)){tempx<-unlist(strsplit(xa[i],"\t"));tempy<- c(tempy, new("TextMining", PMID = as.numeric(tempx[1]), Title = temp[2], Abstract = tempx[3], Journal_Name = tempx[4]))};return(tempy)}

setClass("TextMining", representation(PMID = "numeric", Title ="character",Abstract = "character", Journal_Name = "character"))

Fig. 16. Representation of S4 Class "TextMining" with R scripts to accomplish the construction.

```
S4 Methods
setGeneric("getAbstract", function(object) standardGeneric("getAbstract"));
setMethod("getAbstract","TextMining",function(object){if
(regexpr(pattern="pathway",object@Abstract,fixed=TRUE)!=-1){return
(object@Abstract)}});
```

6.5 Getting data from Pfam and PRINTS for a specified domain or pattern name.

Pfam data for a specified domain name or PRINTS data for specified pattern name may be extracted using S4 scripts. The process is summarized below in form of decision tree (Figure 17). The dataclasses with their representations is described in Figure 18 and Figure 19.

Biological Data Modelling and Scripting in R   223

```
R Scripts
filtered<- sapply(textfile,getAbstract)
doc1 <- filtered[!sapply(filtered,is.null)]
library(tm);library(rJava);library(RWekajars);library(RWeka);library(slam);library(
Snowball);library(corpora);
doc2<-Corpus(VectorSource(doc1))
setcontrol<-list(minDocFreq = 6, removeNumbers = FALSE, stemming = TRUE,
stopwords = TRUE)
TDM<-TermDocumentMatrix(doc2,control=setcontrol)
Terms<-inspect(TDM[1:17,1])
Terms1<-as.data.frame(Terms)
write.table(Terms1,file="Terms1.txt",sep="\t")
write.table(Terms1,file="Terms1.txt",col.names=FALSE,sep="\t")
Terms2<-read.table("Terms1.txt",header=FALSE,sep="\t")
FilteredTerms<-as.character(Terms2[1:17,1])
load("HGNC_GENE_IDS.RData")
load("DrugNamesDrugBank.RData")
AllGeneSymbols<-c(as.character(HGNC_GENE_IDS[1:19363,2])); AllGeneNames<-
c(as.character(HGNC_GENE_IDS[1:19363,3])); AllGeneAliases<-
c(as.character(HGNC_GENE_IDS[1:19363,6])); AllDrugNames<-
c(as.character(DrugNamesDrugBank[1:6824,1]))
getGeneSymbols<-intersect(AllGeneSymbols,FilteredTerms); getGeneNames<-
intersect(AllGeneNames,FilteredTerms); getGeneAliases<-
intersect(AllGeneAliases,FilteredTerms); getDrugNames<-
intersect(AllDrugNames,FilteredTerms)
z1<- charmatch(getGeneSymbols,FilteredTerms)
for(i in 1:4){b<-NULL;b<-z1[i];sumFreqOccurence<-NULL;sumFreqOccurence<-
sum(inspect(TDM[b,1:22]));write(sumFreqOccurence,file="SumFreqOccurence.txt",
append=TRUE,sep=",")}
```

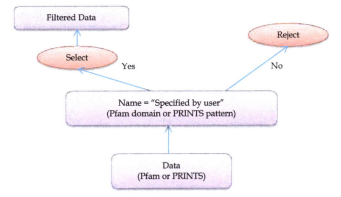

Fig. 17. Decision tree to identify data from Pfam and PRINTS for a specified domain or pattern name.

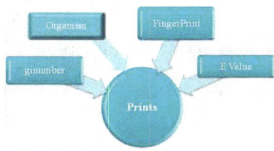

readdata.prints<- function(xz){xa<- readLines(con = xz); tempy<- NULL;for (i in seq (along = xa)){tempx<- unlist(strsplit(xa[i],"\t"));tempy<- c(tempy, new("Prints", ginumber = as.numeric(tempx[1]), organism = tempx[2],FingerPrint=tempx[3], E_value= as.numeric(tempx[4])))};return(tempy)}

setClass("Prints",representation(ginumber = "numeric",organism ="character",FingerPrint="character",E_value= "numeric"))

Fig. 18. Representation of S4 Class "Prints" with R scripts to accomplish the construction.

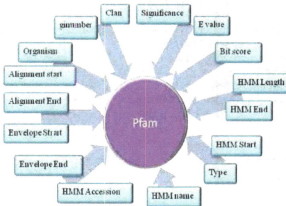

readdata.pfam<- function(xz){xa<- readLines(con = xz); tempy<- NULL;for (i in seq (along = xa)){tempx<- unlist(strsplit(xa[i],"\t"));tempy<- c(tempy, new("Pfam", ginumber = as.numeric(tempx[1]), organism = tempx[2],alignment_start= as.numeric(tempx[3]),alignment_end= as.numeric(tempx[4]),envelope_start= as.numeric(tempx[5]), envelope_end= as.numeric(tempx[6]),hmm_acc= tempx[7],hmm_name= tempx[8],type= tempx[9],hmm_start= as.numeric(tempx[10]),hmm_end= as.numeric(tempx[11]),hmm_length= as.numeric(tempx[12]),bit_score= as.numeric(tempx[13]),E_value= as.numeric(tempx[14]), significance= as.numeric(tempx[15]),clan= tempx[16]))};return(tempy)}

setClass("Pfam",representation(ginumber = "numeric",organism ="character",alignment_start = "numeric",alignment_end = "numeric",envelope_start= "numeric",envelope_end= "numeric",hmm_acc="character",hmm_name="character", type="character",hmm_start= "numeric",hmm_end= "numeric",hmm_length= "numeric",bit_score= "numeric",E_value= "numeric",significance= "numeric",clan="character"))

Fig. 19. Representation of S4 Class "Pfam"along with R scripts to accomplish the construction.

```
S4 Methods
setGeneric("get_prints",function(object,x))
standardGeneric("get_prints"));setMethod("get_prints","Prints",function(object,x){if
( object@FingerPrint == x) {tempo<-
paste(object@ginumber,object@organism,object@FingerPrint,object@E_value, sep="
"); return (tempo)}else {return(0)} })
setGeneric("get_pfam",function(object,x))
standardGeneric("get_pfam"));setMethod("get_pfam","Pfam",function(object,x){if (
object@hmm_name == x) {tempo<- paste(object@ginumber,object@organism,
object@alignment_start, object@alignment_end, object@envelope_start,
object@envelope_end, object@hmm_acc, object@hmm_name, object@type,
object@hmm_start, object@hmm_end, object@hmm_length, object@bit_score,
object@E_value, object@significance, object@clan, sep=" "); return (tempo)}else
{return(0)} })
```

```
R Scripts
result1<- sapply(eclprints,get_prints,"HOMSERKINASE");result1<- setdiff(result1,"0")
result1<- sapply(eclpfam,get_pfam,"DnaJ");result1<- setdiff(result1,"0")
```

7. Parallel computing

When the data size is large (example in millions) and fast information calculation and retrieval is needed, a single modern computational processor fails to fulfill the purpose. In that case, a number of processors are needed to work simultaneously, each carrying out same set of operations on different data objects. This type of approach is called *Parallelization on data level;* the processing time for a single object is not being reduced but a number of data objects are being processed during the same time-interval by separate processors.

The Rmpi package (Yu, 2010) is helpful in this direction. We used ChemmineR (Cao et al 2008) using Rmpi, an interface to MPI (Message Passing Interface). In MPI, processes communicate with each other by sending and receiving messages. We made a library of small molecules of size around 26 millions in SDF file format from various publicly and commercially available sources; these molecules were distributed over 1792 files. We have used ChemmineR's cmp.parse() function to get corresponding atom-pair descriptors and these were stored in .rda files. These .rda files constitute our database. Our aim was to find molecules similar to a given query molecule.

On a typical workstation of 1GB RAM and Intel(R) Pentium(R) 4 CPU 3.40GHz, it takes about 5 hrs to complete a similarity search. In order to increase the speed we have used Rmpi (version: 0.5-8) along with Open MPI (version 1.3.2) – A High Performance Message Passing Library and implemented the Data Level Parallelization on ROCKS (release 4.3) cluster- an open-source Linux cluster distribution with eight nodes having Intel(R) Xeon(TM) CPU 3.60GHz processors. The result was a significant increase in performance and the job was done within 30 minutes. This approach has been shown in Figure 20 as a flowchart.

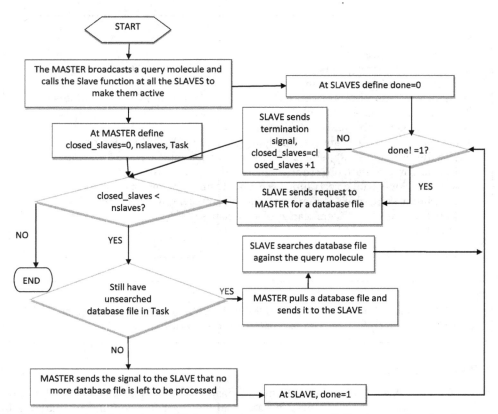

Fig. 20. Figure describing the parallel computing process, parallelization on data level.

8. Building your own package

The various R objects and S4 methods can be encapsulated into .RData named according to the selected problem. All these packages can be sourced from the link http://sourceforge.net/projects/sysbior/.

9. Conclusion

Various packages to handle biological data of diverse type of problems have been developed with the help of S4 object oriented programming. The S4 objects can be handled with ease applying S4 methods to fetch useful data. The process described here may also be modified with novel thought process to create S4 methods to fetch useful analysis data accomplishing other conditions.

10. References

Bui, H.H., Sidney, J., Peters, B., Sathiamurthy, M., Sinichi, A., Purton, K.A., Mothé, B.R., Chisari, F.V., Watkins, D.I. & Sette, A. (2005). Automated generation and evaluation of specific MHC binding predictive tools: ARB matrix applications, *Immunogenetics*, Vol. 57, No. 5, (June 2005), pp. 304-314.

Bruford, E.A., Lush, M.J., Wright, M.W., Sneddon, T.P., Povey, S. & Birney, E. (2008). The HGNC Database in 2008: a resource for the human genome, *Nucleic Acids Res*, Vol. 36, Suppl No. 1, (January 2008), pp. D445-D448.

Cao, Y., Charisi, A., Cheng, L.C., Jiang, T. & Girke, T. (2008). ChemmineR: a compound mining framework for R, *Bioinformatics*, Vol. 24, No. 15, (August 2008), pp. 1733-1774.

Chaudhuri, R., Ahmed, S., Ansari, F.A., Singh, H.V. & Ramachandran, S. (2008). MalVac: database of malarial vaccine candidates, *Malar J*, Vol. 7, No. 184,(September 2008), pp. 1-7.

Chaudhuri, R., Ansari, F.A., Raghunandan, M.V. & Ramachandran, S. (2011). FungalRV: Adhesin prediction and Immunoinformatics portal for human fungal pathogens. *BMC Genomics*, Vol. 12, No. 192, (April 2011).

Cheadle, C., Vawter, M.P. & Freed, W.J. (2003). Analysis of Microarray data using Z-score transformation, *Mol Diagn*, Vol. 5, No. 2, (May 2003), pp. 73-81.

Feinerer, I., Hornik, K. & Meyer, D. (2008). Text mining infrastructure in R, *Journal of Statistical Software*, Vol. 25, No. 5, (March 2008), ISSN 1548-7660.

Fiers, M.W., Kleter, G.A., Nijland, H., Peijnenburg, A.A., Nap, J.P. & van Ham, R.C. (2004). Allermatch, a webtool for the prediction of potential allergenicity according to current FAO/WHO Codex alimentarius guidelines, *BMC Bioinformatics*, Vol. 5, (September 2004).

Gao, Q., Kripke, K.E., Saldanha, A.J., Yan, W., Holmes, S. & Small, P.M. (2005). Gene expression diversity among Mycobacterium tuberculosis clinical isolates, *Microbiology*, Vol. 151, No.1, (January 2005), pp. 5-14.

Gentleman, R.C., Carey, V.J., Bates, D.M., Bolstad, B., Dettling, M., Dudoit, S., Ellis, B., Gautier, L., Ge, Y., Gentry, J., Hornik, K., Hothorn, T., Huber, W., Iacus, S., Irizarry, R., Leisch, F., Li, C., Maechler, M., Rossini, A.J., Sawitzki, G., Smith, C., Smyth, G., Tierney, L., Yang, J.Y. & Zhang, J. (2004). Bioconductor: open software development for computational biology and bioinformatics, *Genome Biol*, Vol. 5, No. 10, (September 2004).

Goodsell, D. S., Morris, G. M. & Olson, A. J. (1996). Automated Docking of Flexible Ligands: Applications of AutoDock, *J. Mol. Recognition*, Vol. 9, pp. 1-5.

Guha, R. (2007). Chemical Informatics Functionality in R, *Journal of Statistical Software*, Vol. 18, No. 5, (January 2007), ISSN 1548-7660.

Hansen, N.T., Kouskoumvekaki, I., Jørgensen, F.S., Brunak, S. & Jónsdóttir, S. O. (2006). Prediction of pH-dependent aqueous solubility of druglike molecules, *J Chem Inf Model*, Vol. 46, No. 6, (November 2006), pp. 2601-2609.

Hess, B., Kutzner, C., van der Spoel, D. & Lindahl, E. (2008). GROMACS 4: Algorithms for Highly Efficient, Load-Balanced, and Scalable Molecular Simulation, *J. Chem. Theory Comput.*, Vol. 4, No. 3, (February 2008), pp. 435–447.

Hornik, K. (August 2009). Snowball: Snowball Stemmers, In: *R-project.org*, 12.12.2010, Available from: http://CRAN.R-project.org/package=Snowball

Hornik, K., Buchts, C. & Zeileis, A. (2009). Open-Source Machine Learning: R Meets Weka, *Computational Statistics*, Vol. 24, No. 2, (May 2009), pp. 225-232, ISSN: 0943-4062.

Irwin, J.J. & Shoichet, B. K. (2005). ZINC--a free database of commercially available compounds for virtual screening, *J Chem Inf Model.*, Vol. 45, No. 1, (2005), pp. 177-182.

Knox, C., Law, V., Jewison, T., Liu, P., Ly, S., Frolkis, A., Pon, A., Banco, K., Mak, C., Neveu, V., Djoumbou, Y., Eisner, R., Guo, A.C. & Wishart, D.S. (2011). DrugBank 3.0: a comprehensive resource for 'omics' research on drugs, *Nucleic Acids Res.*, Vol. 39, Suppl. No. 1, (January 2011), pp. 1035-1041.

Li, P., Zhang, C., Perkins, E., Gong, P. & Deng, Y. (2007). Comparision of probabilistic Boolean network and dyanamic Bayesian network approaches for inferring regulatory networks, *BMC Bioinformatics*, (November 2007), Vol. 8, Suppl. No. 7.

Lundegaard, C., Lamberth, K., Harndahl, M., Buus, S., Lund, O. & Nielsen, M. (2008). NetMHC-3.0: accurate web accessible predictions of human, mouse and monkey MHC class I affinities for peptides of length 8-11, *Nucleic Acids Res.*, Vol. 36, Suppl. No. 2, pp. W509-W512.

Maimon, O. & Rokach, L. (2005). Decision Trees, In: Data Mining and Knowledge Discovery Handbook, O. Maimon, & L. Rokach, (Ed.), pp. 165-192, Springer, ISBN: 978-0-387-25465-4, United States of America.

MarvinSketch 5.3.8. (2010). In: ChemAxon, 10.11.2010, Available from: http://www.chemaxon.com

Müssel, C., Hopfensitz, M. & Kestler, H.A. (2010). BoolNet--an R package for generation, reconstruction and analysis of Boolean networks, Bioinformatics, Vol. 26, No. 10, (May 2010), pp. 1378-1380

Nelson, D.L. & Cox, M.M. (2000). Enzymes, In: Lehninger Principles of biochemistry, 3rd Edition, W.H. Freeman, (Ed.), pp. 190-237, Worth Publishers, ISBN 1-57259-9316, New York.

Parker, K.C., Bednarek, M.A. & Coligan, J.E. (1994). Scheme for ranking potential HLA-A2 binding peptides based on independent binding of individual peptide side-chains, J Immunol., Vol. 152, No. 1, (January 1994), pp. 163-175, ISSN: 0022-1767.

R Development Core Team (2010). R: A language and environment for statistical computing, In: R Foundation for Statistical Computing, ISBN 3-900051-07-0, Available from: http://www.R-project.orgSaha, S. & Raghava, G.P.S. (2006). AlgPred: prediction of allergenic proteins and mapping of IgE epitopes, *Nucleic Acids Res.*, Vol. 34, Suppl. No. 2, (July 2006), pp. W202-209.

Saha, S. & Raghava, G.P.S. (2006). Prediction of Continuous B-cell Epitopes in an Antigen Using Recurrent Neural Network, *Proteins*, Vol. 65, No. 1, (October 2006), pp. 40-48.

Saha, S. & Raghava, G. P. S. (2007). Prediction methods for B-cell epitopes, *Methods Mol. Biol.*, Vol. 409, No. 4, pp. 387-394.

Singh, H. & Raghava, G. P. S. (2001). ProPred: Prediction of HLA-DR binding sites, *Bioinformatics*, Vol. 17, No. 12, (December 2001), pp. 1236-1237.

Tyson, J., Chen, K. & Novak, B. (2001). Network dynamics and cell physiology, *Nature Rev. Mol Cell. Biol.*, Vol. 2, No. 12, (December 2001), pp. 908-916.

Viswanathan, G., Seto, J., Patil, S., Nudelman, G. & Sealfon, S. (2008). Getting started in biological pathway construction and analysis, *PLoS comp. biol.*, Vol. 4, No. 2 (e16), (February 2008), pp. 0001-0005.

Vivona, S., Gardy, J. L., Ramachandran, S., Brinkman, F.S., Raghava, G.P.S, Flower, D.R. & Filippini, F. (2008). Computer-aided biotechnology: from immuno-informatics to reverse vaccinology, *Trends Biotechnol.*, Vol. 26, No. 4, (April 2008), pp. 190-200.

Voigt, J.H., Bienfait, B., Wang, S. & Nicklaus, M.C. (2001). Comparison of the NCI open database with seven large chemical structural databases, *J Chem Inf Comput Sci.*, Vol. 41, No. 3, (May-Jun 2001), pp. 702-712.

Yu, H. (November 2010). Rmpi: Interface (Wrapper) to MPI (Message-Passing Interface), In: *cran.r-project.org*, 6.1.2011, Available from: http://CRAN.R-project.org/package=Rmpi

Zhang, Q., Wang, P., Kim, Y., Haste-Andersen, P., Beaver, J., Bourne, P. E., Bui, H. H., Buus, S., Frankild, S., Greenbaum, J., Lund, O., Lundegaard, C., Nielsen, M., Ponomarenko, J., Sette, A., Zhu, Z. & Peters, B. (2008). Immune epitope database analysis resource (IEDB-AR), *Nucleic Acids Res.*, Vol. 36, Suppl. No. 2, (July 2008), pp. W513-W518.

Permissions

All chapters in this book were first published in SCBBCM, by InTech Open; hereby published with permission under the Creative Commons Attribution License or equivalent. Every chapter published in this book has been scrutinized by our experts. Their significance has been extensively debated. The topics covered herein carry significant findings which will fuel the growth of the discipline. They may even be implemented as practical applications or may be referred to as a beginning point for another development.

The contributors of this book come from diverse backgrounds, making this book a truly international effort. This book will bring forth new frontiers with its revolutionizing research information and detailed analysis of the nascent developments around the world.

We would like to thank all the contributing authors for lending their expertise to make the book truly unique. They have played a crucial role in the development of this book. Without their invaluable contributions this book wouldn't have been possible. They have made vital efforts to compile up to date information on the varied aspects of this subject to make this book a valuable addition to the collection of many professionals and students.

This book was conceptualized with the vision of imparting up-to-date information and advanced data in this field. To ensure the same, a matchless editorial board was set up. Every individual on the board went through rigorous rounds of assessment to prove their worth. After which they invested a large part of their time researching and compiling the most relevant data for our readers.

The editorial board has been involved in producing this book since its inception. They have spent rigorous hours researching and exploring the diverse topics which have resulted in the successful publishing of this book. They have passed on their knowledge of decades through this book. To expedite this challenging task, the publisher supported the team at every step. A small team of assistant editors was also appointed to further simplify the editing procedure and attain best results for the readers.

Apart from the editorial board, the designing team has also invested a significant amount of their time in understanding the subject and creating the most relevant covers. They scrutinized every image to scout for the most suitable representation of the subject and create an appropriate cover for the book.

The publishing team has been an ardent support to the editorial, designing and production team. Their endless efforts to recruit the best for this project, has resulted in the accomplishment of this book. They are a veteran in the field of academics and their pool of knowledge is as vast as their experience in printing. Their expertise and guidance has proved useful at every step. Their uncompromising quality standards have made this book an exceptional effort. Their encouragement from time to time has been an inspiration for everyone.

The publisher and the editorial board hope that this book will prove to be a valuable piece of knowledge for researchers, students, practitioners and scholars across the globe.

List of Contributors

Simone Cristoni
Ion Source Biotechnologies srl, Milano

Silvia Mazzuca
Plant Cell Physiology laboratory, Università della Calabria, Rende, Italy

Gamil Abdel-Azim
College of Computer, Qassim University, Saudi Arabia
College of Computer & Informatics, Canal Suez University, Egypt

Aboubekeur Hamdi-Cherif
College of Computer, Qassim University, Saudi Arabia
Computer Science Department, Université Ferhat Abbas, Setif (UFAS), Algeria

Mohamed Ben Othman
College of Computer, Qassim University, Saudi Arabia
The research Unit of Technologies of Information and Communication (UTIC) / ESSTT, Tunisia

Z. A. Aboeleneen
College of Computer& Informatics, Zagazig University, Egypt

James J. Cai
Texas A&M University, College Station, Texas, USA

Jestin Jean-Luc and Lafaye Pierre
Institut Pasteur France

Pascal Kahlem et al.
EMBL -European Bioinformatics Institute, Wellcome Trust Genome Campus, Hinxton, Cambridge, United Kingdom

Robert Stewart and Yonggang Zhu
CSIRO Materials Science and Engineering Australia

Iftah Gideoni
CSIRO Information and Communication Technology Centre, Australia

Geraldine Sandana Mala John and Chellan Rose
Central Leather Research Institute, India

Satoru Takeuchi
Factory of Takeuchi Nenshi ,Takenen Japan

José Carlos Jiménez-López, María Isabel Rodríguez-García and Juan de Dios Alché
Department of Biochemistry, Cell and Molecular Biology of Plants, Estación Experimental del Zaidín, CSIC, Granada, Spain

Di Liu and Juncai Ma
Network Information Center, Institute of Microbiology, Chinese Academy of Sciences WFCC-MIRCEN World Data Centre for Microorganisms (WDCM) China, People's Republic

Srinivasan Ramachandran et al.
G.N. Ramachandran Knowledge Centre for Genome Informatics, Institute of Genomics and Integrative Biology, Delhi, India

Index

A
Affinity Maturation, 74, 77-78, 84-86, 88
Allergens, 163-164, 173-178, 203
Allergy, 163, 173, 176-178
Amino Acid Sequence, 135, 137-138, 152, 164, 178
Amino Acids, 25, 33, 51, 53, 56, 80, 83, 136-138, 140, 189, 203
Antibody, 68-79, 82, 84-88, 174, 177-178
Antibody Fragments, 70-72, 76, 79, 84-86, 88
Antigen, 69-79, 83-84, 86-87, 203, 228
Attractors, 106, 214, 216-217

B
B Cell, 73, 206, 208
Barcode Of Life Data System, 180, 185-186, 188, 199
Baseline Noise, 112-115, 117-122, 124, 126, 129
Bioinformatics, 1-2, 11, 18-19, 21-23, 27, 41-44, 48, 65, 82, 89, 158-159, 182-184, 190, 195-197, 201-203, 211, 227-228
Bioinformatics Tools, 1, 136, 154-155, 196, 202, 211
Biological Systems, 2, 6, 90, 93, 95-96, 144, 210-211
Boolean Network, 95, 99, 104, 106, 210, 214, 228

C
Chemoinformatics, 217-218
Clustering Algorithm, 151, 190
Codon, 51, 54-55, 57, 66, 71, 88, 189

D
Decision Tree, 99, 101, 202, 204, 207-208, 213, 218-219, 221-223
Denoising, 116, 119-120, 127, 129-131, 133

Descriptor, 28, 30-34, 40, 175, 220
Dna Barcoding, 179-186, 194, 196-200

E
Electropherogram, 111-114, 116-117, 119-120, 122, 125
Enzyme, 18, 68, 71, 79-82, 93-94, 104, 211
Epitopes, 75-77, 83, 88, 203-204, 208, 228
Evolutionary Bioinformatics, 43, 48, 65
Evolutionary Distance, 50-54, 151

G
Gene Expression, 3, 5-6, 9, 20, 26, 90, 93, 97, 99, 108-109, 212, 227
Gene Regulatory Network, 109, 210
Genomics, 2-4, 7, 18-20, 22, 24-26, 104, 106, 151-152, 164, 173, 181, 198, 201, 227
Genotype, 67-68, 70, 90

H
Haplotypes, 44-45, 60, 62, 65
Helices, 135, 139-140, 143, 150, 156-157, 159-160
Hellinger Distance, 28, 30-32, 34-37, 40

I
Immune Response, 73, 75, 202-203
Immunoinformatics, 203-204, 227

L
Ligand, 71, 139-140, 218
Local Similarity, 150, 154

M
Mass Spectrometry, 2, 9, 11, 14-15, 19-26, 117, 119-121, 127, 130-131, 133, 135

Matlab, 35, 37, 40, 43-50, 52-53, 55, 57, 59, 64-65, 211

Metabolic Pathway, 90-91

Molecular Biology, 2-3, 23, 26, 40, 65, 82, 84-88, 90, 132, 145, 163, 175, 177-178, 196-197

Multiple Sequence Alignment, 27, 40-42, 83, 148, 150, 156-157, 178

Mutagenesis, 26, 77-80, 85-86, 88

Mutation, 17-18, 59, 62, 66, 73-74, 77-78, 84, 86, 138

N

Nucleic Acids, 24, 40, 69, 71, 104, 106-109, 132, 151, 175-178, 197-198, 227-228

O

Ole E 1 Family, 163-164, 171-173

Ole E 1 Protein, 164-171, 174

P

Parallel Processing, 49, 67, 73, 79-83

Pattern Matching, 111-112, 125-126, 128, 132

Peak Detection, 117, 119-121, 124-129, 131-133

Peak Extraction, 112-113, 116, 122, 124, 126

Peptide, 11-15, 17-18, 22, 24, 69-70, 86, 88, 128, 136-140, 142, 159-160, 174-175, 228

Peptide Bond, 137, 140, 142

Phage, 68-70, 76-79, 81, 84-88, 106

Phenotype, 6, 67-68, 70, 90, 102-103, 105

Phylogenetic Tree, 46, 57, 165, 171

Pollen Protein, 163, 172

Polypeptide, 84, 135-140, 142-143, 155, 163

Polypeptide Chain, 136-140, 142-143, 163

Primary Structure, 135-137, 139, 159-160

Probability Density Function, 30-31, 40

Prosite, 148-149, 160, 165, 176, 178

Protein Analysis, 1, 13, 135-136, 159, 165

Protein Database, 7, 9, 22, 145, 150, 171, 175

Protein Sequences, 19, 33, 50-51, 54, 66, 80, 83, 135, 144-145, 150, 156, 159, 165, 203, 218

Protein Structure, 101, 108, 135-137, 141, 144, 147, 149, 152-155, 157-162, 173-174, 202, 218

Proteomics, 1-2, 4, 6-7, 9-11, 15, 17-26, 127, 129-130, 159, 165, 173, 175-176

Q

Query Sequence, 148, 150, 152, 160

R

R Scripts, 204, 206-208, 210-212, 214, 219-220, 222-225

Recombination, 47, 62, 73-74, 76, 85, 181

Regulatory Networks, 96, 98, 105-106, 108, 214, 228

Ribosome, 67, 70-71, 77-79, 85-86, 88

S

Secondary Structure, 135-136, 138-139, 145, 150-151, 155-160, 162

Secondary Structure Prediction, 136, 156-157, 159-160, 162

Sequence Alignment, 27, 29, 40-42, 58, 83, 136, 148-150, 155-157, 176, 178, 190

Sequence Information, 14, 135-136, 144-145, 148-149, 161, 192

Signal Processing, 43, 111-113, 116-118, 122, 124-126, 129-130, 132-133

T

T Cell, 109, 203, 206, 208

Tertiary Structure, 135, 139-140, 155, 159

Threading, 152, 154-155

Transcription, 5-7, 89, 97, 138, 214

Transcriptome, 5, 20-22, 25-26, 90